T0205534

Studies in Systems, Decision and Control

Volume 471

Series Editor

Janusz Kacprzyk, Systems Research Institute, Polish Academy of Sciences, Warsaw, Poland

The series "Studies in Systems, Decision and Control" (SSDC) covers both new developments and advances, as well as the state of the art, in the various areas of broadly perceived systems, decision making and control–quickly, up to date and with a high quality. The intent is to cover the theory, applications, and perspectives on the state of the art and future developments relevant to systems, decision making, control, complex processes and related areas, as embedded in the fields of engineering, computer science, physics, economics, social and life sciences, as well as the paradigms and methodologies behind them. The series contains monographs, textbooks, lecture notes and edited volumes in systems, decision making and control spanning the areas of Cyber-Physical Systems, Autonomous Systems, Sensor Networks, Control Systems, Energy Systems, Automotive Systems, Biological Systems, Vehicular Networking and Connected Vehicles, Aerospace Systems, Automation, Manufacturing, Smart Grids, Nonlinear Systems, Power Systems, Robotics, Social Systems, Economic Systems and other. Of particular value to both the contributors and the readership are the short publication timeframe and the world-wide distribution and exposure which enable both a wide and rapid dissemination of research output.

Indexed by SCOPUS, DBLP, WTI Frankfurt eG, zbMATH, SCImago.

All books published in the series are submitted for consideration in Web of Science.

Mehdi Soltanifar · Hamid Sharafi ·
Farhad Hosseinzadeh Lotfi · Witold Pedrycz ·
Tofigh Allahviranloo

Preferential Voting and Applications: Approaches Based on Data Envelopment Analysis

Springer

Mehdi Soltanifar
Department of Mathematics, Semnan
Branch
Islamic Azad University
Semnan, Iran

Farhad Hosseinzadeh Lotfi
Department of Mathematics, Science
and Research Branch
Islamic Azad University
Tehran, Iran

Tofigh Allahviranloo
Faculty of Engineering and Natural
Sciences
Istinye University
Istanbul, Türkiye

Hamid Sharafi
Department of Mathematics, Science
and Research Branch
Islamic Azad University
Tehran, Iran

Witold Pedrycz
Department of Electronics
and Communications Engineering
University of Alberta
Edmonton, AB, Canada

ISSN 2198-4182   ISSN 2198-4190 (electronic)
Studies in Systems, Decision and Control
ISBN 978-3-031-30405-7   ISBN 978-3-031-30403-3 (eBook)
https://doi.org/10.1007/978-3-031-30403-3

This Springer imprint is published by the registered company Springer Nature Switzerland AG
The registered company address is: Gewerbestrasse 11, 6330 Cham, Switzerland

Preface

The issue of "choice" using the aggregation of voters' votes is one of the most important group decision-making issues that are always considered by decision makers in electoral systems. Voting is a method of group decision-making in a democratic society that expresses the will of the majority. Voting is perhaps the simplest way to gather the opinions of experts, and this ease of application has made it a multi-attribute decision-making method in group decisions. Preferential voting is a type of voting that may refer to electoral systems or groups of the electoral system. In preferential voting, voters vote for multiple candidates, and how the candidates are arranged on the ballot is important. Researchers have made many efforts to provide models of voter aggregation, and one of the best results of these efforts is the aggregation of votes based on the policy of data envelopment analysis. Thus, in group decisions, the opinions of experts are obtained in a simple structure and consolidated in an interactive and logical structure, and the results can be a powerful tool for decision support.

The purpose of this book is to present the theory and application of the models presented in this regard and to establish a meaningful relationship between data envelopment analysis and multi-attribute decision-making. We intend to introduce these methods as a powerful way in group decision support by providing a complete set of voting models based on data envelopment analysis and expressing its various applications in industry and society. However, most decision-making methods do not use the opinions of experts or reduce the motivation of experts to participate in complex interactions and time, while voting methods do not have this shortcoming.

This book is suitable for graduate students in the fields of industrial management, business management, industrial engineering, applied mathematics and economics. It can also be a good source for researchers in decision science, decision support systems, data envelopment analysis, supply chain management, healthcare management and others. The methods presented in this book can not only offer a comprehensive framework for solving the problems of these areas but also can inspire researchers to pursue new innovative hybrid methods.

We gratefully acknowledge those who have contributed to the compilation of this book, and it is hoped that this book would be useful for readers, researchers and managers.

Semnan, Iran   Mehdi Soltanifar
Tehran, Iran   Hamid Sharafi
Tehran, Iran   Farhad Hosseinzadeh Lotfi
Edmonton, Canada   Witold Pedrycz
Istanbul, Türkiye   Tofigh Allahviranloo
May 2023

Contents

Chapter 1
Basic Concepts of Voting

Abstract The problem of "election" using the aggregation of voters' votes is one of the most important problems of group decision-making for which several models have been presented so far. Consider a group of people who need to make a group decision based on voters' votes. Given that people have different opinions and tastes, how can they make a decision that includes all individual opinions? From such a simple issue to the choice of politicians, all of them require a collective decision. In this chapter, this concept and methods of aggregating individual votes to reach a collective decision are discussed.

1.1 Introduction

In this chapter, the basic concepts of preferential voting as one of the methods of group decision-making are discussed. Using the opinions of a group of experts will always have more satisfactory results in decision making. Therefore, methods based on group opinions are usually more powerful tools for decision support. How to gather the opinions of the group is a very important issue in these methods, which has a direct impact on the final results. In this chapter, it will be discussed about how to aggregate experts' opinions in a voting process. The issue of "choice" using the aggregation of voters' votes is one of the most important issues of group decision-making, which is always the focus of decision-makers in electoral systems. Voting is perhaps the easiest way to gather experts' opinions, and this ease of use has made it a multi-attribute decision-making method in group decision-making. Researchers have made many efforts to provide voter aggregation models, and one of the best results of these efforts is vote aggregation based on data envelopment analysis policy. Therefore, in group decision-making, experts' opinions are obtained in a simple structure and aggregated in an interactive and logical structure, and the results can be a powerful tool to support decision-making. In most decision-making methods, experts' opinions are not used, or in complex and time-consuming interactions, the motivation of experts to participate is reduced. While voting methods do not have this shortcoming. Therefore, it is necessary to study how to aggregate voters' votes in order to achieve a favorable decision-making method.

1.2 Voting Methods

Decision-making is one of the most important issues that every human being faces during his life. The results of a decision sometimes have such an impact on people's lives that it can change the course of life. Therefore, researchers have conducted many studies on decision-making, which has led to several methods. Methods based on group opinions always have more satisfactory results in decision making. A fundamental problem that any group faces is how to reach a good group decision when there is disagreement among its members. Consider a group of people who need to make a group decision. Considering that people have different opinions and tastes, and each of them looks at the issue from a different point of view and expertise, how can they make a decision that takes into account all opinions, expertise, and points of view? The result of these decisions in cases such as choosing a politician can have a significant impact on social life.

Voting is a method of aggregating individual votes to reach a group decision. Voting theory determines how voters' votes are aggregated. Human desire for democracy and participation of society members in determining the governance structure is increasing every day. The struggle for the right to vote and the choice of the future of a country by the people of that country has been a difficult and sometimes bloody challenge in the world for at least the last two centuries. This right and its importance in human bio-political history is enough to be the first answer to the question "Why should we vote?" and this answer is the use of an important and perhaps sacred right. The most reliable, easiest and least expensive way to announce voice and views for decision making; it is the ballot box and this proposition has been proven in many countries of the world at least during the last century.

Contrary to popular belief, participation in elections is not only limited to electing politicians. When a committee has to choose a candidate for employment, when a movie or a football player is chosen, or even when a group of friends has to decide where to go for dinner, an election has actually taken place. Therefore, people always participate in many elections. Therefore, not only important decisions such as parliamentary decisions, approval of bills, etc. need an electoral system, but this system is also seen in simple everyday decisions. In fact, making a decision, choosing a candidate, etc. by means of voting is reasonable and intelligent because the individual opinions of all people share in it. Mathematicians, philosophers, political scientists, and economists have devised various voting methods that select a winner (or winners) from a set of options, taking everyone's opinion into account. It is not difficult to find examples where different voting methods select different winners given the same input from group members. Determining the criteria for comparing different voting methods is not only an interesting and difficult theoretical research, but also has important practical implications; because it will lead to choosing the appropriate method of aggregating votes to achieve satisfactory results.

The ballot structure are generally in two forms. Categorical and ordinal lines. In the first form, voters vote for a single candidate, but in the second form, voters vote for more than one candidate, which are divided into two subgroups: one is that only the names of several candidates are written on the ballot, and the second is that in addition to selecting several candidates, voters also rank their preferences (non-preference and preference) [1]. In the same way, voting is done in both direct and indirect ways. In the first way, the voters choose their candidate directly, and in the second way, the voters first choose people, and the best candidate is determined according to the opinions of the selected people.

In the mechanism of direct and non-preference voting, voters choose r ($r < n$) candidates from among n candidates. Therefore, each voter votes for at most r candidates on the ballot, and the candidate who gets the most votes is the winner (k). One of the disadvantages of this voting, which one of the most common is voting methods, is that voting preferences are not taken into account, and therefore voters cannot convey their preferences to the society.

In some voting methods, voters can select predetermined ballots that have the closest preference to their own opinions. In contrast to this method and in another method, each voter can freely state his preferences for candidates on the ballot. Therefore, another attitude that should be considered in the voting process is the type of voters' choice, which can be strategic or freely (sincerely). In the sincerely mode, the voter completes a ballot according to his preferences regarding the prioritization of the candidates. In the strategic mode of the voter, according to the information he has about how other voters choose, he chooses a ballot that leads to the most favorable result for him (her) [2].

Other voting methods include negative voting. In this type of voting, each voter can support one or more candidates with $+1$ points or reject one or more candidates with -1 points. In this voting system, the winner is the candidate who has the most points. Other voting methods that need to be discussed are approval and cumulative voting. In the approval voting method, voters choose a subset of candidates with the ability to abstain [3]. In the cumulative voting method, the voters are asked to allocate a fixed number of points between the candidates, and the candidate who gets the most points will win [4].

1.3 Preferential Voting: Motivation and Application Necessity

In preferential voting, more information is obtained from voters than other electoral systems. In this type of voting, the voters are not only asked to choose the candidate they want, but also they are asked to introduce the second candidate who is the most worthy if the first choice does not win. Likewise, if their first and second choices

Table 1.1 Vote counting results version 1

Candidates	The number of votes in different priorities			
	First	Second	Third	Fourth
A	7	12	0	0
B	6	0	3	10
C	6	4	0	9
D	0	3	16	0

do not win, their third choice will be requested [5]. Therefore, in the preferential voting system, each voter chooses a subset of candidates and ranks them based on their preferences, and then the best candidate is determined by the totality of voters' votes. There are different ways to choose the best candidate.

Suppose there is a group of 19 people who want to choose a representative from among 4 candidates A, B, C and D. Also assume that the results of the vote counting are shown in the Table 1.1. Voters have chosen candidates in an electoral system based on preferential voting with 4 priorities.

Due to the difference of opinions of the voters, there is usually no candidate that is of interest to all the voters. If there were two candidates, it would be easier and the winner could be chosen by the candidate who supports more than 50% of the voters. It should be mentioned that even in this case, the above sentence can be interpreted in different ways. Therefore, there are different attitudes to choose the winning candidate. At first glance at the Table 1.1, it seems that candidate A is the best candidate; because this candidate has been prioritized more than other candidates. On the other hand, more than half of the voters ranked candidate A last, which shows that there is a difference of opinion regarding candidate A. Candidate D has different conditions, this candidate has no votes in the first and second priority. This issue cannot rule out the possibility of candidate D winning; but if we pay more attention to the ballots in Table 1.2, where candidate D is always listed after candidate B in the ballots; maybe this issue cannot be a reason why candidate B won, but it shows that candidate D should definitely not win the election; because candidate B is better than him (her).

According to the above explanations, the selection of the winning candidate can be limited to candidate B and C. This issue was raised in the eighth century between

Table 1.2 Vote counting results version 2

Number of voters	Ballot template			
	First priority	Second priority	Third priority	Fourth priority
6	C	B	D	A
6	B	C	D	A
4	A	B	D	C
3	A	C	B	D

Borda [6] and Condorcet [7] as the first scientific theorists in the field of voting methods. According to Condorcet, candidate B is the winner. He stated that the winning candidate is the candidate who can get the majority of votes in pairwise comparisons between candidates. According to Condorcet's attitude to select the winning candidate, it is clear that: the difference between the better rank of candidate C and other candidates is 5, 1 and 11 respectively for candidates A, B and D. While this issue is for candidate B with other candidates 5 and 19 for candidates A and D respectively. But candidate B was 10 times higher than candidate C and candidate C was 9 times higher than candidate B. So candidate B will win.

In plurality rule, the voter chooses a candidate and the candidate with the most votes is the winner. This method is also called "First Past the Post". Despite its widespread use, this method has serious problems. One of them is not transferring voters' preferences to society. For more explanation, consider Table 1.1. It is assumed that each voter's vote is its first priority, so candidate A will win with 7 votes. While we discussed earlier and saw that candidate A is a Condorcet loser. After that, candidate B and candidate C are with 6 votes. If it can be seen, this method often cannot transfer the majority of the society. Of course, some suggestions have been made to solve this problem, the most famous of which is the setting of a certain threshold to select the winning candidate.

In quota rule, the process is like plurality rule and the winning candidate is the candidate who has at least $q \times N$ votes. q is the assigned score between zero and one, and N is the total number of voters. Majority rule, is another method which is actually in this method $q = 0.5$. If $q = 1$, it is called unanimity rule. One of the major problems of this method is the absence of a candidate who has the minimum number of votes. In other words, in many cases, this method cannot determine the winning candidate. In the voting example mentioned in Table 1.1, considering the first vote of each voter and placing $q = 0.5$, no candidate will win. Although no method is the best, some methods are superior to others. A very popular category of vote aggregation methods is the priority weighting method, in which fixed weights are assigned to different priorities. Among these methods, Borda count and scoring rule can be mentioned. In these methods, predetermined weights are assigned to the votes in each priority, and the candidate with the most scores will be the winner. Obviously, the weight given to the votes in the first priority should not be less than the weight given to the second priority, and in the same way, the weights of the following priorities are also limited. Let v_{rj} represent the number of rth$(r = 1, \ldots, k)$ priority votes for the jth$(j = 1, \ldots, n)$ candidate. The Overall Desirability Index (ODI) for each candidate is defined as follows.

$$Z_j = \sum_{r=1}^{k} w_r v_{rj}. \tag{1.1}$$

Borda [6] proposed the weights as $w_r = k - r + 1 \; r = 1, \ldots, k$ and specified that the winning candidate is the candidate who has the highest ODI, and thus in the example of Table 1.1, candidate B with ODI $= 60$ is the winner.

$$Z_A = 4 \times 7 + 3 \times 0 + 2 \times 0 + 1 \times 12 = 40$$
$$Z_B = 4 \times 6 + 3 \times 10 + 2 \times 3 + 1 \times 0 = 60$$
$$Z_C = 4 \times 6 + 3 \times 9 + 2 \times 4 + 1 \times 0 = 57$$
$$Z_D = 4 \times 0 + 3 \times 0 + 2 \times 16 + 1 \times 3 = 36$$

The above method is a special case of the scoring rule, in which a sequence of fixed weights is assigned to the number of votes in each priority, and the candidate with the most ODI will win; so that the weights have the order $w_1 \geq w_2 \geq ... \geq w_k$. Based on this, plurality rule can be considered a special mode of scoring rule in which the weight of the first priority is one and the weight of the rest of the priorities will be zero. In other words, $w_1 = 1$, $w_r = 0$ $(r = 2, ..., k)$. To express another perspective of scoring rule, we can refer to the k-approval voting method. k-approval voting is a method in which each voter chooses k number of candidates and the winner is the candidate with the most votes (in this method, no ranking or prioritization is done by the voter). Looking more closely at the issue, if the priorities 1 to k are assigned a weight of one in the scoring rule and the rest of the priorities have zero weight, the k-Approval voting method is obtained. According to the example mentioned in Table 1.1, the winner of k-approval voting with $k = 1$, is candidate A; with $k = 2$ are candidates B and C; and with $k = 3$ is candidate C. Obviously, the above may happen in many votings. In other words, one of the problems of the scoring rule method is the absence of a unique winning candidate, as well as how to determine the weights, which will definitely be sensitive to the selection of the winning candidate. In other words, the selection of predetermined weights does not necessarily maximize the ODI for each candidate, and it changes with a different weight vector of the winning candidate, and certainly causes the candidates to protest. In the same way, other methods such as elimination procedure can be used [8].

1.4 The Voting Paradox

Suppose there are three voters and 3 candidates a, b and c, whose preferences are as follows:

$$Voter\ 1 : a \succ b,\ b \succ c\ and\ a \succ c$$
$$Voter\ 2 : b \succ c,\ c \succ a\ and\ b \succ a$$
$$Voter\ 3 : c \succ a,\ a \succ b\ and\ c \succ b$$

Two voters preferred b to c, also two voters preferred a to b, so it seems reasonable that the majority preferred a to c, while the opposite is true for the second and third voters. Therefore, it cannot be said that the majority preferred a to c. This conclusion can be expressed for the rest of the candidates. According to what was explained above, it is clear that none of the candidates has a preference over the other, and therefore, none of the candidates can be announced as the winner. The above event in the voting process is known as the voting paradox [9]. In the voting process with more than two voters, each voter has his (her) individual preferences. The main problem is to aggregate individual preferations and obtain a unique group ranking. The technique of aggregating votes is called social welfare function. Arrow [10] by expressing the intervals of comparability of candidates and establishing superiority between candidates, introduced the following five conditions:

Condition 1: Universal Admissibility of Individual Orderings.
Condition 2: Non-Perversity or Positive Association of Individual and Social Values.
Condition 3: Independence from Irrelevant Alternatives.
Condition 4: Citizens' Sovereignty.
Condition 5: Non-dictatorship.

Then, Arrow [10] stated in a theorem that any social welfare function that holds the principles of comparability and multiplicity and is true in conditions 1 to 3, will violate condition 4 or condition 5. Davis et al. [11] mentioned a formulation different from Arrow and used probabilistic calculations for the social selection problem. Yu [12] introduced a class of solutions for the group decision problem using the concept of the ideal solution (utopia point), and finally Cook and Kress [13] presented a ranking using the distance function and Similar to Arrow [10], they introduced a set of principles that every distance function should apply to.

1.5 Discussion and Example

In this section, an important feature for the selection methods of the best candidate is stated and discussed with an example.

It was mentioned earlier that one of the ways to use the Plurality Rule is to set a threshold limit for selecting the winning candidate; in fact, the candidate who won at least 50% of the votes. If there is no such candidate, the method used is plurality with runoff, in which, except for the first and second candidates who are evaluated by the plurality rule method, the rest of the candidates are eliminated and the criterion of obtaining at least 50% of the votes is checked again to select the winning candidate. The developed mode of this method is the hare rule, which is the same as the plurality with runoff method; with the difference that in each stage the candidate with the lowest rank is eliminated and the criterion of obtaining at least 50% of votes is checked to select the winning candidate from among the remaining candidates. In case of not getting at least 50% of the votes, the last candidate will be eliminated and the criteria of winning will be checked. Obviously, several candidates may win together in the end.

One of the issues raised by researchers in the field of social studies and voting is that the methods of choosing the best candidate should have uniformity or alignment. In other words, the rise of a candidate in the ratings made by the voters should not

Table 1.3 Counting votes in two scenarios

Number of voters	Ballot template	
	The first scenario	The second scenario
6	ABCD	ABCD
4	BCAD	BCAD
5	CABD	CABD
2	BACD	ABCD
2	DCAB	DCAB

have a negative effect on their election and reduce the chance of the candidate being elected. With a simple look, it is clear that plurality rule has this feature and it is shown here that plurality with runoff does not have this feature. The subject of Borda's method is also discussed. For this purpose, consider the following example, in which 18 voters have ranked candidates *A*, *B*, *C*, and *D* according to their preferences, and the results of counting votes are displayed in the Table 1.3.

The first point that should be mentioned about the Table 1.3 is that both scenarios are the same, with the difference that they have two different ballots, which is evident in the fourth line of the scenarios in the table. Actually, in the first scenario, candidate *B* has performed better and in the second scenario, candidate *A* has performed better. Now, the steps to determine the winning candidate in both scenarios using the plurality with runoff method are shown in Table 1.4.

In the Table 1.4 and in the first scenario, after checking the number of votes of the candidates in the first step, the last two candidates (*C* and *D*), are eliminated, and by eliminating them, candidate *A* has been ranked first 13 times and therefore, this candidate is the winner. In the second scenario, after eliminating candidates *B* and *D*, candidate *C* has been in the first place 11 times and is therefore the winner. As can be seen in the Table 1.3, although in the second scenario, candidate *A* has two more votes in the first place, but with the plurality with runoff method, candidate *A* did not win, while in the first scenario he (she) had won. In other words, as the votes of candidate *A* increased, the chances of this candidate decreased in the plurality with runoff method, and this method does not have a uniform feature. A similar example can be used for the hare rule method and show that this method also does not have uniformity; but there is another point about the hare rule method that needs to be discussed here. Considering the first scenario and using the hare rule method, as well

Table 1.4 The process of selecting the winning candidate (The plurality with runoff method)

The first scenario						The second scenario					
Step	Eliminated candidate	Number of votes				Step	Eliminated candidate	Number of votes			
		A	B	C	D			A	B	C	D
1	–	6	6	5	2	1	–	8	4	5	2
2	C, D	13	6	0	0	2	B, D	8	0	11	0
–	–	Winner	–	–	–	–	–	–	–	Winner	–

Table 1.5 The process of selecting the winning candidate (The hare rule method)

The first scenario with the elimination of A in the second step						The first scenario with the elimination of B in the second step					
Step	Eliminated candidate	Number of votes				Step	Eliminated candidate	Number of votes			
		A	B	C	D			A	B	C	D
1	–	6	6	5	2	1	–	6	6	5	2
2	D	6	6	7	0	2	D	6	6	7	0
3	A	0	12	7	0	3	B	8	0	11	0
–	–	–	Winner	–	–	–	–	–	–	Winner	–

Table 1.6 The process of selecting the winning candidate (The Borda's method)

Candidate	ODI in the first scenario	ODI in the second scenario
A	57	59
B	54	52
C	54	54
D	25	25

as eliminating the candidate who has the least votes, two identical choices emerge. Candidates A and B both have 6 votes, and if each is elected, the winning candidate will be different.

The results of the hare rule method can be seen in Table 1.5. In this method, by eliminating candidate B, candidate C will win, and by eliminating candidate A, candidate B will win. This is one of the disadvantages of this method, which is caused by not taking into account all the preferences of the voters. In general, as stated earlier, considering the number of votes in the first place will remove a lot of information from the issue.

Now by Borda's method for the data in Table 1.3, the score of each candidate in each scenario will be as Table 1.6. This method as well as the Condorcet method have the property of uniformity, and the improvement of a candidate's condition will not reduce his (her) chances of winning.

1.6 Summary

There are several methods to determine the mechanism of voting systems, which were summarized in this chapter. In fact, in this chapter, a process was presented that specifies the need to discuss and review preferential voting methods. Voting is a social phenomenon that can have different perspectives and as a result has multiple interpretations. Naturally, there are different voting methods and newer methods may be added to them. The important point is that by studying different voting methods, one can understand the type of use of voting systems from voters' opinions. In many cases, voting systems try to increase the flexibility of voters to increase their ability to vote and choose. These efforts include abstention, negative voting, scoring systems and preferential voting. It is obvious that the main goal of all electoral and voting systems is firstly to make a collective decision and secondly to make a decision that the society supports. So, the more appropriate the mechanism designed to record voters' opinions, the more satisfied the society will be. Therefore, in this chapter, voting, especially preferential voting, was introduced as one of the group decision-making methods for decision support.

References

1. Reynolds, A., Reilly, B., Ellis, A., Cheibub, J.: Electoral System Design: The New International IDEA Handbook. International Institute for Democracy and Electoral Assistance, Political Science (2005)
2. Hansson, S., Grüne-Yanoff, T.: Preferences, the stanford encyclopedia of philosophy. In: Zalta, E. (ed.) Metaphysics Research Lab, Stanford University (2009)
3. Laslier, J., Sanver, M. (eds.): Springer, Berlin, Heidelberg (2010)
4. Balinski, M., Laraki, R.: A theory of measuring, electing and ranking. Proc. Natl. Acad. Sci. **104**(21), 8720–8725 (2007)
5. Reilly, B.: The global spread of preferential voting: Australian institutional imperialism? Aust. J. Polit. Sci. **39**(2), 253–266 (2004)
6. Borda, J.: Mémoire sur les élections au scrutin. Histoire de L'Académie Royale des Sciences **102**, 657–665 (1781)
7. Condorcet, M.D.: Essai sur l'application de l'analyse à la probabilité des décisions, rendues à la pluralité des voix/par M. le Marquis de Condorcet..., a Paris: De l'imprimerie Royale (1785)
8. Straffin, P.: Topics in the Theory of Voting. Birkhauser, Boston (1980)
9. Riker, W.: Voting and summation of preferences: an interpretive bibliographical review of selected developments during the last decade. Am. Polit. Sci. Rev. **5**, 900–911 (1961)
10. Arrow, K.: Social Choice and Individual Values. Wiley, New York (1951)
11. Davis, O., Degroot, M., Hinchi, M.: Social preference orderings and majority rule. Econometrica **40**(1), 147–157 (1972)
12. Yu, P.: A class of solutions for group decision problems. Manage. Sci. **19**(8), 936–946 (1973)
13. Cook, W., Kress, M.: Priority ranking and consensus formation. Manage. Sci. **24**, 1721–1732 (1978)

Chapter 2
Introduction to Data Envelopment Analysis

Abstract Efficiency is a management concept that has a long history in management science. Efficiency shows that an organization has used its resources in a good way in order to produce the best performance at a point in time. One of the appropriate and efficient tools in the field of efficiency measurement and evaluation is Data Envelopment Analysis (DEA), which is used as a non-parametric method to calculate the efficiency of decision-making units. In fact, DEA is based on a series of optimizations using linear programming, which is used to evaluate the efficiency of Decision-Making Units (DMUs) that have multiple inputs and multiple outputs. This method is based on an optimistic policy that can evaluate DMUs in the best conditions. In this chapter, the basic methods and basic principles of DEA are discussed.

2.1 Introduction

Evaluating the performance of an organization's subdivisions has always been the focus of the organization's managers. A technique that can evaluate units with several criteria so that it can identify their strengths and weaknesses, can have many applications in most organizations. This chapter introduces a technique to achieve this purpose. This technique, called Data Envelopment Analysis (DEA), evaluates the performance of a set of homogeneous Decision-Making Units (DMUs). Identify all the strengths and weaknesses of each DMU in each of the criteria by comparing the DMUs. This technique has made significant progress in recent decades and has had many researches in modeling and application [1–10].

In this Chapter, basic DEA models are designed using subject principles and the relationship between them will be discussed. Then the weight restrictions and their addition to the models will be analyzed and the problems that may arise in these models as well as the solutions to these models will be presented.

© The Author(s), under exclusive license to Springer Nature Switzerland AG 2023   11
M. Soltanifar et al., *Preferential Voting and Applications: Approaches Based on Data Envelopment Analysis*, Studies in Systems, Decision and Control 471,
https://doi.org/10.1007/978-3-031-30403-3_2

2.2 Production Possibility Set (PPS) and Basic DEA Models

In this section, with the help of the principles of the subject, first the PPS is introduced and then the basic DEA models are designed. For this purpose, suppose there are n homogeneous DMUs. The DMU_j uses the input vector $\mathbf{X_j} = \left(x_{1j}, \ldots, x_{mj}\right)$ to generate the output vector $\mathbf{Y_j} = \left(y_{1j}, \ldots, y_{sj}\right)$ such that, $\mathbf{X_j} \geq \mathbf{0}, \mathbf{X_j} \neq \mathbf{0}; \mathbf{Y_j} \geq \mathbf{0}, \mathbf{Y_j} \neq \mathbf{0}(j = 1, 2, \ldots, n)$.

Definition 2.1 (*DEA Technique*): A technique based on mathematical programming to evaluate the performance of a set of homogeneous DMUs with multiple inputs and multiple outputs.

Based on the definition of DEA technique for each DMU in the set of observations, and comparing the inputs and outputs of that unit with other existing and observed units, performance evaluation is performed and its strengths and weaknesses are identified [11].

Definition 2.2 (*PPS*): A set of units (*T*) in which outputs can be obtained from inputs [12]

$$T = \left\{ \begin{pmatrix} \mathbf{X} \\ \mathbf{Y} \end{pmatrix} \middle| \mathbf{X} \text{ can produce } \mathbf{Y} \right\}$$

The PPS includes all the observed DMUs in the community as well as some unseen units that could have existed. Unseen units in a community can be identified by existing rules and properties. For this purpose, the following principles are considered to make the PPS. It should be noted that in a community, some of the following principles may be established and true, and some may not be. Determining which of the principles of the subject is established in a community depends on the laws governing that community, which establish the acceptance or non-acceptance of the principles. In fact, the principles in a community are unproven basic rules.

Principle 1 (Inclusion of Observations)
$$\forall j \left((1 \leq j \leq n) \Rightarrow \begin{pmatrix} \mathbf{X_j} \\ \mathbf{Y_j} \end{pmatrix} \in T \right)$$

Principle 2 (Convexity) According to this principle, the PPS is convex.

$$\forall \begin{pmatrix} \mathbf{X'} \\ \mathbf{Y'} \end{pmatrix} \forall \begin{pmatrix} \mathbf{X''} \\ \mathbf{Y''} \end{pmatrix} \forall \lambda \left[\left(\begin{pmatrix} \mathbf{X'} \\ \mathbf{Y'} \end{pmatrix} \& \begin{pmatrix} \mathbf{X''} \\ \mathbf{Y''} \end{pmatrix} \in T \& \lambda \in [0, 1] \Rightarrow \lambda \begin{pmatrix} \mathbf{X'} \\ \mathbf{Y'} \end{pmatrix} + (1 - \lambda) \begin{pmatrix} \mathbf{X''} \\ \mathbf{Y''} \end{pmatrix} \in T \right]$$

Principle 3 (Return to Scale) This principle can be presented in the following three ways:

Constant Return to Scale (CRS): $\forall \begin{pmatrix} \mathbf{X} \\ \mathbf{Y} \end{pmatrix} \left(\begin{pmatrix} \mathbf{X} \\ \mathbf{Y} \end{pmatrix} \in T \& \lambda \geq 0 \Rightarrow \begin{pmatrix} \lambda \mathbf{X} \\ \lambda \mathbf{Y} \end{pmatrix} \in T \right)$.

Increase Return to Scale (IRS): $\forall \begin{pmatrix} \mathbf{X} \\ \mathbf{Y} \end{pmatrix} \left(\begin{pmatrix} \mathbf{X} \\ \mathbf{Y} \end{pmatrix} \in T \ \& \ \lambda \geq 1 \Rightarrow \begin{pmatrix} \lambda \mathbf{X} \\ \lambda \mathbf{Y} \end{pmatrix} \in T \right).$

Decrease Return to Scale (DRS): $\forall \begin{pmatrix} \mathbf{X} \\ \mathbf{Y} \end{pmatrix} \left(\begin{pmatrix} \mathbf{X} \\ \mathbf{Y} \end{pmatrix} \in T \ \& \ 0 \leq \lambda \leq 1 \Rightarrow \begin{pmatrix} \lambda \mathbf{X} \\ \lambda \mathbf{Y} \end{pmatrix} \in T \right).$

Principle 4 (Possibility) This principle can be presented in the following two ways:

Inputs possibility: $\forall \begin{pmatrix} \mathbf{X} \\ \mathbf{Y} \end{pmatrix} \left(\begin{pmatrix} \mathbf{X} \\ \mathbf{Y} \end{pmatrix} \in T \ \& \ \overline{\mathbf{X}} \geq \mathbf{X} \Rightarrow \begin{pmatrix} \overline{\mathbf{X}} \\ \mathbf{Y} \end{pmatrix} \in T \right).$

Outputs possibility: $\forall \begin{pmatrix} \mathbf{X} \\ \mathbf{Y} \end{pmatrix} \left(\begin{pmatrix} \mathbf{X} \\ \mathbf{Y} \end{pmatrix} \in T \ \& \ \overline{\mathbf{Y}} \leq \mathbf{Y} \Rightarrow \begin{pmatrix} \mathbf{X} \\ \overline{\mathbf{Y}} \end{pmatrix} \in T \right).$

Principle 5 (Minimum Interpolation) T is the smallest set that applies to the selected principles.

In addition to the above principles, other principles may be defined in a community. Not all of these principles may apply to a community. The first principle (inclusion of observations) is established in any community because observations come from the same society. The fifth principle (minimum interpolation) guarantees the uniqueness of PPS and also ensures that at least one of the DMUs is on the efficiency frontier and will therefore be relatively efficient. Therefore, acceptance of Principle 5 is mandatory.

Acceptance or non-acceptance of other principles depends on the community. The following are some of the special PPSs that have been considered by most researchers.

2.2.1 The CCR Model

By accepting the principles "inclusion of observations", "convexity", "CRS", "possibility" and "minimum interpolation", the following PPS are created [13]:

$$T_C = \left\{ \begin{pmatrix} \mathbf{X} \\ \mathbf{Y} \end{pmatrix} \middle| \mathbf{X} \geq \sum_{j=1}^{n} \lambda_j \mathbf{X_j} \ \& \ \mathbf{Y} \leq \sum_{j=1}^{n} \lambda_j \mathbf{Y_j} \ \& \ \lambda_j \geq 0 (j = 1, \ldots, n) \right\} \quad (2.1)$$

T_C in the space of one input and one output is shown in Fig. 2.1.

Definition 2.3 DMU $\begin{pmatrix} \mathbf{X_k} \\ \mathbf{Y_k} \end{pmatrix}$ is better than DMU $\begin{pmatrix} \mathbf{X_l} \\ \mathbf{Y_l} \end{pmatrix}$ if and only if $\begin{pmatrix} -\mathbf{X_k} \\ \mathbf{Y_k} \end{pmatrix} \overset{\geq}{\neq} \begin{pmatrix} -\mathbf{X_l} \\ \mathbf{Y_l} \end{pmatrix}.$

Fig. 2.1 T_C in the space of
one input-one output

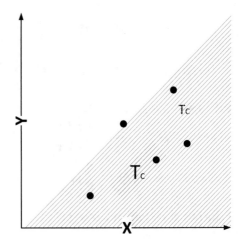

Definition 2.4 DMU$_p$ in a PPS such as T is called relative efficient if and only if no better unit than DMU$_p$ is found in T.

There are several ways to answer the question: "Is there a better unit in PPS than DMUp?" There are three ways to check for better units in the PPS for the unit under evaluation.

i.   Input-oriented: Improving inputs (reducing inputs)
ii. Output-oriented: Improved output (increased output)
iii. Combined-oriented: Improving inputs and outputs (reducing inputs and increasing outputs).

Different models in DEA are designed based on the choice of how to improve and also with the acceptance or non-acceptance of different principles.

If T_C is considered, the CCR model is designed [14]. This model will be in the input-oriented as follows:

$$\text{Min } \theta$$

$$\text{s.t. } \begin{pmatrix} \theta \mathbf{X_p} \\ \mathbf{Y_p} \end{pmatrix} \in T_C \tag{2.2}$$

This model can be converted to model (2.3) based on the properties of the set (2.1).

$$\text{Min } \theta$$

$$s.t. \sum_{j=1}^{n} \lambda_j \mathbf{X_j} \leq \theta \mathbf{X_p}$$

$$\sum_{j=1}^{n} \lambda_j \mathbf{Y_j} \geq \mathbf{Y_p}$$

$$\lambda_j \geq 0. \tag{2.3}$$

In model (2.3) by selecting $\theta = 1$, $\lambda_p = 1$ and $\lambda_j = 0$, $j \neq p$, a feasible solution is obtained. Hence $0 < \theta^* \leq 1$.

Obviously, if $\theta^* < 1$, then at T_C a unit with less input than $\mathbf{X_p}$ and with the same output level $\mathbf{Y_p}$ is found. In this case, DMU$_p$ is not efficient. But if $\theta^* = 1$, then in T_C it is not possible to find a better unit than DMU$_p$ by reducing all the inputs to a same scale. Of course, it may be possible to find a better unitt in the T_C by reducing some of the components of input vector $\mathbf{X_p}$ and with the same output level $\mathbf{Y_p}$. Therefore if $\theta^* = 1$ then it cannot be conclusively concluded that DMU$_p$ is a relatively efficient. The following model is designed to address this issue.

$$\text{Min } Z_p = \theta - \varepsilon \left(\sum_{i=1}^{m} s_i^- + \sum_{r=1}^{s} s_r^+ \right)$$

$$s.t. \ \sum_{j=1}^{n} \lambda_j x_{ij} + s_i^- = \theta x_{ip}, \ i = 1, \ldots m$$

$$\sum_{j=1}^{n} \lambda_j y_{rj} - s_r^+ = y_{rp}, \ r = 1, \ldots, s$$

$$\lambda_j \geq 0, \ j = 1, \ldots, n$$

$$s_i^- \geq 0, \ i = 1, \ldots, m$$

$$s_r^+ \geq 0, \ r = 1, \ldots, s \tag{2.4}$$

In model (2.4) if $s_i^{-*} = 0 (i = 1, \ldots, m)$, $s_r^{+*} = 0 (r = 1, \ldots, s)$ and $\theta^* = 1$ or in other words $Z_p^* = 1$, then DMU$_p$ is relatively efficient; otherwise DMU$_p$ is inefficient. In this model, ε is a very small positive value.

The output-oriented CCR model is as follows:

$$\text{Max } \varphi$$

$$s.t. \ \begin{pmatrix} \mathbf{X_p} \\ \varphi \mathbf{Y_p} \end{pmatrix} \in T_C \tag{2.5}$$

Model (2.5) can be converted to model (2.6) based on the properties of the set (2.1).

$$\text{Max } \varphi$$

$$s.t. \ \sum_{j=1}^{n} \lambda_j \mathbf{X_j} \leq \mathbf{X_p}$$

$$\sum_{j=1}^{n} \lambda_j \mathbf{Y_j} \geq \varphi \mathbf{Y_p}$$

$$\lambda_j \geq 0. \tag{2.6}$$

Assuming that φ^* is the optimal value of the model (2.6), $\varphi^* \geq 1$. If $\varphi^* > 1$ then DMU_p will be inefficient. Also if $\varphi^* = 1$ then it can not be concluded that DMU_p is relatively efficient. Similar to the input-oriented model, the output-oriented CCR model for distinguishing efficient from inefficient units can be presented as follows:

$$\text{Max } W_p = \varphi + \varepsilon \left(\sum_{i=1}^{m} s_i^- + \sum_{r=1}^{s} s_r^+ \right)$$

$$s.t. \ \sum_{j=1}^{n} \lambda_j x_{ij} + s_i^- = x_{ip}, \ i = 1, ...m$$

$$\sum_{j=1}^{n} \lambda_j y_{rj} - s_r^+ = \varphi y_{rp}, \ r = 1, \ldots, s$$

$$\lambda_j \geq 0, \ j = 1, \ldots, n$$

$$s_i^- \geq 0, \ i = 1, \ldots, m$$

$$s_r^+ \geq 0, \ r = 1, \ldots, s \tag{2.7}$$

In model (2.7) if $W_p^* = 1$ then DMU_p is relatively efficient; otherwise DMU_p is inefficient. Value θ^* in model (2.4) and value $\frac{1}{\varphi^*}$ in model (2.7) are called relative efficiency of DMU_p, which holds in Eq. (2.8):

$$\frac{1}{\varphi^*} = \theta^* \tag{2.8}$$

2.2.2   The BCC Model

By accepting the principles of "inclusion of observations", "convexity", "possibility" and "minimum interpolation", the following PPS is obtained.

$$T_V = \left\{ \begin{pmatrix} \mathbf{X} \\ \mathbf{Y} \end{pmatrix} \middle| \mathbf{X} \geq \sum_{j=1}^{n} \lambda_j \mathbf{X_j}, \mathbf{Y} \leq \sum_{j=1}^{n} \lambda_j \mathbf{Y_j}, \sum_{j=1}^{n} \lambda_j = 1, \lambda_j \geq 0 (j = 1, ..., n) \right\} \tag{2.9}$$

In fact, T_V has been obtained by removing the CRS principle from T_C, and some call it the PPS with Variable Returns to Scale (VRS). The input-oriented BCC model is as follows [15]:

$$\text{Min } \alpha_p = \theta - \varepsilon \left(\sum_{i=1}^{m} s_i^- + \sum_{r=1}^{s} s_r^+ \right)$$

$$s.t. \sum_{j=1}^{n} \lambda_j x_{ij} + s_i^- = \theta x_{ip}, \quad i = 1, \ldots m;$$

$$\sum_{j=1}^{n} \lambda_j y_{rj} - s_r^+ = y_{rp}, \quad r = 1, \ldots, s;$$

$$\sum_{j=1}^{n} \lambda_j = 1;$$

$$\lambda_j \geq 0(j = 1, \ldots, n); s_i^- \geq 0(i = 1, \ldots, m); s_r^+ \geq 0(r = 1, \ldots, s); \quad (2.10)$$

The difference between the BCC and CCR models is in the presence of constraint $\sum_{j=1}^{n} \lambda_j = 1$, which has an impact on how the target is found in the PPS.

If $\alpha_p^* = 1$ then DMU$_p$ is efficient in T_V, otherwise DMU$_p$ is inefficient. Since T_V has one less principle than T_C, so $T_C \subset T_V$; also according to the models (2.4) and (2.10), if θ_{CCR}^* and θ_{BCC}^* are the relative efficiency of DMU$_p$ obtained from models (2.4) and (2.10) then $\theta_{BCC}^* \geq \theta_{CCR}^*$. Therefore, if DMU$_p$ is relatively efficient in T_C, then it is also relatively efficient in T_V.

The output-oriented BCC model is as follows:

$$\text{Max } \beta_p = \varphi + \varepsilon \left(\sum_{i=1}^{m} s_i^- + \sum_{r=1}^{s} s_r^+ \right)$$

$$s.t. \sum_{j=1}^{n} \lambda_j x_{ij} + s_i^- = x_{ip}, \ i = 1, \ldots m$$

$$\sum_{j=1}^{n} \lambda_j y_{rj} - s_r^+ = \varphi y_{rp}, \ r = 1, \ldots, s$$

$$\sum_{j=1}^{n} \lambda_j = 1;$$

$$\lambda_j \geq 0, \ j = 1, \ldots, n$$
$$s_i^- \geq 0, \ i = 1, \ldots, m$$
$$s_r^+ \geq 0, \ r = 1, \ldots, s \quad (2.11)$$

If $\beta_p^* = 1$ then DMU$_p$ is efficient in T_V, otherwise DMU$_p$ is inefficient.

Definition 2.5 The strong efficiency frontier of the PPS T is defined as follows:

Fig. 2.2 T_V in the space of one input-one output

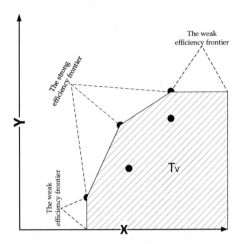

$$\partial_S T = \left\{ \begin{pmatrix} \mathbf{X} \\ \mathbf{Y} \end{pmatrix} \in T \left| \left(\begin{pmatrix} \overline{\mathbf{X}} \\ \overline{\mathbf{Y}} \end{pmatrix} \in T, \begin{pmatrix} -\overline{\mathbf{X}} \\ \overline{\mathbf{Y}} \end{pmatrix} \geq \begin{pmatrix} -\mathbf{X} \\ \mathbf{Y} \end{pmatrix} \right) \Rightarrow \overline{\mathbf{X}} = \mathbf{X} \& \overline{\mathbf{Y}} = \mathbf{Y} \right. \right\}$$

(2.12)

In fact, the strong efficiency frontier is a set of PPS points that are not dominated by any other point in the PPS.

Definition 2.6 The weak efficiency frontier of the PPS T is defined as follows:

$$\partial_W T = \sim \left[\exists \begin{pmatrix} \overline{\mathbf{X}} \\ \overline{\mathbf{Y}} \end{pmatrix} \left(\begin{pmatrix} \mathbf{X} \\ \mathbf{Y} \end{pmatrix} \in \mathrm{T} \& \begin{pmatrix} -\overline{\mathbf{X}} \\ \overline{\mathbf{Y}} \end{pmatrix} > \begin{pmatrix} -\mathbf{X} \\ \mathbf{Y} \end{pmatrix} \right) \right]$$

(2.13)

Figure 2.2 shows the T_V with its strong and weak efficiency frontiers.

2.2.3   The FDH Model

By accepting the principles of "inclusion of observations", "possibility" and "minimum interpolation", the following PPS is obtained.

$$T_{FDH} = \left\{ \begin{pmatrix} \mathbf{X} \\ \mathbf{Y} \end{pmatrix} \middle| \mathbf{X} \geq \sum_{j=1}^{n} \lambda_j \mathbf{X_j}, \mathbf{Y} \leq \sum_{j=1}^{n} \lambda_j \mathbf{Y_j}, \right.$$

$$\left. \sum_{j=1}^{n} \lambda_j = 1, \lambda_j \in \{0, 1\} (j = 1, \ldots, n) \right\}$$

(2.14)

T_{FDH} can also be expressed as follows:

$$T_{FDH} = \bigcup_{j=1}^{n} \left\{ \begin{pmatrix} X \\ Y \end{pmatrix} \middle| X \geq X_j \& Y \leq Y_j \right\} \tag{2.15}$$

Based on the properties of this PPS, input-oriented is FDH model as follows [16]:

$$\text{Min } \theta$$

$$s.t. \sum_{j=1}^{n} \lambda_j x_{ij} \leq \theta x_{ip}, \ i = 1, \ldots m$$

$$\sum_{j=1}^{n} \lambda_j y_{rj} \geq y_{rp}, \ r = 1, \ldots, s$$

$$\sum_{j=1}^{n} \lambda_j = 1$$

$$\lambda_j \in \{0, 1\}, \ j = 1, \ldots, n \tag{2.16}$$

Model (2.16) is a mixed integer programming problem, so it has a high computational complexity. Using the properties of PPS (2.15), the model (2.16) can also be expressed as follows:

$$\theta_j = \text{Max } \theta$$

$$s.t. \ X_j \leq \theta X_p$$

$$Y_j \geq Y_p \tag{2.17}$$

Model (2.17) must be solved for $j \in \{1, \ldots, n\}$. If the model is feasible, θ_j is obtained from the following equation:

$$\theta_j = \text{Max} \left\{ \frac{x_{ij}}{x_{ip}} \middle| x_{ip} > 0 \right\} \tag{2.18}$$

Model (2.17) will not have a feasible solution in the following conditions.

$$\forall j \left(Y_j \underset{\neq}{\geq} Y_p \right) \vee \exists i (x_{ip} = 0 \& x_{ij} > 0) \tag{2.19}$$

After solving model (2.17) for $j = 1, 2, \ldots, n$, the optimal solution is obtained from the following equation:

$$\theta_p^* = \text{Min} \left\{ \theta_j \middle| \text{model (2.17) is feasible for } j \right\} \tag{2.20}$$

The output-oriented FDH model, like the input-oriented model, can be converted to n optimization problems and can be easily solved with a process with low computational complexity.

2.2.4   The CHD Model

By accepting the principles of "inclusion of observations", "convexity" and "minimum interpolation", the following PPS is obtained.

$$T_{CHD} = \left\{ \begin{pmatrix} \mathbf{X} \\ \mathbf{Y} \end{pmatrix} \middle| \mathbf{X} = \sum_{j=1}^{n} \lambda_j \mathbf{X_j}, \ \mathbf{Y} = \sum_{j=1}^{n} \lambda_j \mathbf{Y_j}, \right.$$
$$\left. \sum_{j=1}^{n} \lambda_j = 1, \lambda_j \geq 0 (j = 1, \dots, n) \right\} \tag{2.21}$$

T_{CHD} is actually the Convex Hull of DMUs (CHD). This PPS is shown in Fig. 2.3 [17].

In T_{CHD}, non-radial models should be used to evaluate the units. The modified Russell model for evaluating DMU_p in this PPS can be presented as follows:

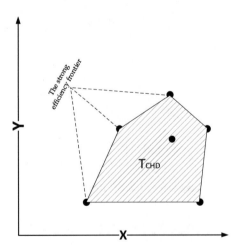

Fig. 2.3 T_{CHD} in the space of one input-one output

$$\text{Min} \quad \frac{\frac{1}{m}\sum_{i=1}^{m}\theta_i}{\frac{1}{s}\sum_{r=1}^{s}\varphi_r}$$

$$s.t. \quad \sum_{j=1}^{n}\lambda_j x_{ij} = \theta_i x_{ip}, \; i = 1, \ldots m$$

$$\sum_{j=1}^{n}\lambda_j y_{rj} = \varphi_r y_{rp}, \; r = 1, \ldots, s$$

$$\sum_{j=1}^{n}\lambda_j = 1$$

$$\theta_i \leq 1, \; i = 1, \ldots m$$

$$\varphi_r \geq 1, \; r = 1, \ldots, s$$

$$\lambda_j \geq 0, \; j = 1, \ldots, n \tag{2.22}$$

Assuming $\mathbf{X_p} > \mathbf{0}$ and $\mathbf{Y_p} > \mathbf{0}$ in model (2.22) and replacing relations (2.23), this model becomes model (2.24) which is known as SBM model [18].

$$\begin{cases} \forall i \; \theta_i = \frac{x_{ip}-s_i^-}{x_{ip}} \\ \forall r \; \varphi_r = \frac{y_{rp}+s_r^+}{y_{rp}} \end{cases} \tag{2.23}$$

$$\text{Min} \; \rho_p = \frac{1 - \frac{1}{m}\sum_{i=1}^{m}\frac{s_i^-}{x_{ip}}}{1 + \frac{1}{s}\sum_{r=1}^{s}\frac{s_r^+}{y_{rp}}}$$

$$s.t. \quad \sum_{j=1}^{n}\lambda_j x_{ij} = x_{ip} - s_i^-, \; i = 1, \ldots m$$

$$\sum_{j=1}^{n}\lambda_j y_{rj} = y_{rp} + s_r^+, \; r = 1, \ldots, s$$

$$\sum_{j=1}^{n}\lambda_j = 1$$

$$\mathbf{S^-} \geq \mathbf{0}, \; \mathbf{S^+} \geq \mathbf{0}$$

$$\lambda_j \geq 0, \; j = 1, \ldots, n \tag{2.24}$$

If $\rho_p^* = 1$ then DMU_p is efficient in T_{CHD}, otherwise DMU_p is inefficient. The value of ρ_p^* is called the relative efficiency of DMU_p.

2.3   The Concept of Relative Efficiency

In this section, basic models of DEA with the help of weights are presented. There-fore, after expressing the concept of absolute efficiency and relative efficiency, models will be presented. Assuming that each DMU has an input (x_j) and an output (y_j), the absolute performance of DMU_p is obtained from equation $E_p = \frac{y_p}{x_p}$. The relative efficiency of DMU_p is also calculated from the following equation.

$$RE_p = \frac{E_p}{\max\{E_j, j = 1, \ldots, n\}} = \frac{\frac{y_p}{x_p}}{\max\left\{\frac{y_j}{x_j}, j = 1, \ldots, n\right\}} \tag{2.25}$$

If for every j, $(x_j, y_j) > \mathbf{0}$ then $0 < RE_p \leq 1$. Also $RE_p = 1$ if and only if the absolute efficiency of DMU_p is the maximum value of the absolute efficiency of the observation set.

Now suppose DMU_j has input vector $\mathbf{X_j} \in \mathbb{R}^m_{\geq 0}$ and output vector $\mathbf{Y_j} \in \mathbb{R}^s_{\geq 0}$. Developing the definition of absolute and relative efficiency of one input and one output to multiple inputs and multiple outputs requires considering the weight for inputs and outputs. Assume that the weight vector $\mathbf{U} = (u_1, \ldots, u_s)$ is corresponding to the output vector $\mathbf{Y_j} \in \mathbb{R}^s_{\geq 0}$ and the weight vector $\mathbf{V} = (v_1, \ldots, v_m)$ corresponds to the input vector $\mathbf{X_j} \in \mathbb{R}^m_{\geq 0}$, so that $\mathbf{U} > \mathbf{0}, \mathbf{V} > \mathbf{0}$.

Definition 2.7 DMU_p is called efficient if and only if

$$\exists \mathbf{U} \, \exists \mathbf{V} \, \forall j \left((\mathbf{U}, \mathbf{V}) > \mathbf{0} \& \frac{\mathbf{UY_p}}{\mathbf{VX_p}} \geq \frac{\mathbf{UY_j}}{\mathbf{VX_j}} \right) \tag{2.26}$$

The following model is presented to determine whether DMU_p is efficient or inefficient [19].

$$\text{Max } RE_p = \frac{\frac{\mathbf{UY_p}}{\mathbf{VX_p}}}{\max_{1 \leq j \leq n}\left\{\frac{\mathbf{UY_j}}{\mathbf{VX_j}}\right\}}$$

$$s.t. \ (\mathbf{U}, \mathbf{V}) > \mathbf{0} \tag{2.27}$$

In model (2.27) we can replace $(\mathbf{U}, \mathbf{V}) \geq \left(\underbrace{\varepsilon, \ldots, \varepsilon}_{(s + m) \text{ times}} \right)$ with $(\mathbf{U}, \mathbf{V}) > \mathbf{0}$, where ε is a small non-Archimedean positive number. If $RE_p = 1$ then DMU_p is efficient, otherwise DMU_p is inefficient. The value of RE_p is called the relative efficiency of DMU_p. Model (2.27) can be converted as follows by applying Charnes and Cooper transformations [20]:

$$\text{Max } \frac{\mathbf{UY_p}}{\mathbf{VX_p}}$$

$$s.t. \frac{\mathbf{UY_j}}{\mathbf{VX_j}} \leq 1, j = 1, \ldots, n$$

$$(\mathbf{U}, \mathbf{V}) \geq \mathbf{1}\varepsilon. \qquad (2.28)$$

Model (2.28) is known as the CCR fractional model and is converted to the following form using the Charnes and Cooper transformations:

$$\text{Max } \mathbf{UY_p}$$

$$s.t. \ \mathbf{VX_p} = 1$$

$$\mathbf{UY_j} - \mathbf{VX_j} \leq 0, j = 1, \ldots, n$$

$$(\mathbf{U}, \mathbf{V}) \geq \mathbf{1}\varepsilon \qquad (2.29)$$

The extended form of the above model is as a model (2.30):

$$\text{Max } \sum_{r=1}^{s} u_r y_{rp}$$

$$s.t. \sum_{i=1}^{m} v_i x_{ip} = 1$$

$$\sum_{r=1}^{s} u_r y_{rj} - \sum_{i=1}^{m} v_i x_{ij} \leq 0, j = 1, \ldots, n$$

$$u_i \geq \varepsilon, i = 1, \ldots, m$$

$$v_r \geq \varepsilon, r = 1, \ldots, s \qquad (2.30)$$

It can be easily shown that model (2.30) is actually dual of model (2.4). Therefore, the value of efficiency calculated by models obtained from multiplicative method and subject principles method, are equal. The only difference is that in the multiplicative model, weights are calculated for each of the inputs and outputs, with the help of which the defining hyperplanes of the PPS can be obtained. From the solution of the cover models, the coordinates of the target point are calculated and with the help of it, the reference set can also be obtained.

The multiplicative model for calculating relative efficiency without accepting the principle of CRS is as follows:

$$\text{Max } RE_p = \frac{\frac{\mathbf{UY_p} + u_0}{\mathbf{VX_p}}}{\max_{1 \leq j \leq n} \left\{ \frac{\mathbf{UY_j} + u_0}{\mathbf{VX_j}} \right\}}$$

$$s.t. \ (\mathbf{U}, \mathbf{V}) \geq \mathbf{1}\varepsilon. \qquad (2.31)$$

Model (2.31) can be converted as follows by applying Charnes and Cooper transformations:

$$\text{Max } \frac{\mathbf{UY_p} + u_0}{\mathbf{VX_p}}$$
$$s.t. \frac{\mathbf{UY_j} + u_0}{\mathbf{VX_j}} \leq 1, j = 1, \ldots, n$$
$$(\mathbf{U}, \mathbf{V}) \geq \mathbf{1}\varepsilon \tag{2.32}$$

This model becomes the following linear programming model by reusing the Charnes and Cooper transformations.

$$\text{Max } \mathbf{UY_p} + u_0$$
$$s.t. \mathbf{VX_p} = 1,$$
$$\mathbf{UY_j} + u_0 - \mathbf{VX_j} \leq 0, j = 1, \ldots, n$$
$$(\mathbf{U}, \mathbf{V}) \geq \mathbf{1}\varepsilon. \tag{2.33}$$

It can be easily shown that model (2.33) is actually dual of model (2.11). Therefore, the value of efficiency calculated by models obtained from multiplicative method and subject principles method, are equal.

2.4 Models Without Explicit Inputs or Outputs

In some application issues, there are modes where all evaluation indicators are input or all output. Therefore, the DMU has no explicit inputs or outputs [21].

Suppose the DMU has only an output vector and no input [22]. In this case, the relative efficiency of DMU_p in multiplicative form is obtained by solving the following model.

$$\text{Max } \frac{\mathbf{UY_p}}{\max_{1 \leq j \leq n} \{\mathbf{UY_j}\}}$$
$$s.t. \mathbf{U} \geq \mathbf{1}\varepsilon. \tag{2.34}$$

This model becomes the following linear programming model by reusing the Charnes and Cooper transformations.

$$\text{Max } \mathbf{UY_p}$$
$$s.t. \mathbf{UY_j} \leq 1, j = 1, \ldots, n$$
$$\mathbf{U} \geq \mathbf{1}\varepsilon \tag{2.35}$$

The objective function of the model (2.35) can be considered as $\frac{UY_p}{1}$ and its constraints can be considered as $\frac{UY_j}{1} \leq 1$. This means that model (2.35) for DMUs without explicit inputs is equivalent to a fixed input value equale 1, for all DMUs. The dual model (2.35) is as follows:

$$\text{Min} \sum_{j=1}^{n} \lambda_j$$

$$s.t. \sum_{j=1}^{n} \lambda_j Y_j \geq Y_p$$

$$\lambda_j \geq 0, \, j = 1, \ldots, n \tag{2.36}$$

By placing $\sum_{j=1}^{n} \lambda_j = \theta,$, the model (2.36) can be considered as follows.

$$\text{Min} \; \theta$$

$$s.t. \sum_{j=1}^{n} \lambda_j = \theta$$

$$\sum_{j=1}^{n} \lambda_j Y_j \geq Y_p$$

$$\lambda_j \geq 0, \, j = 1, \ldots, n \tag{2.37}$$

This model can also be converted to model (2.38).

$$\text{Min} \; \theta$$

$$s.t. \sum_{j=1}^{n} \lambda_j \times 1 \leq \theta \times 1$$

$$\sum_{j=1}^{n} \lambda_j Y_j \geq Y_p$$

$$\lambda_j \geq 0, \, j = 1, \ldots, n \tag{2.38}$$

The first constraint in the model (2.38) indicates that for all DMUs a fixed input equal to 1 is considered.

The output-oriented model for the mode without explicit inputs is model (2.39).

$$\text{Max} \; \varphi$$

$$s.t. \sum_{j=1}^{n} \lambda_j Y_j \geq \varphi Y_p$$

$$\lambda_j \geq 0, \, j = 1, \ldots, n \tag{2.39}$$

In model (2.39) we have $\varphi \rightarrow +\infty$, if $\lambda_j \rightarrow +\infty$, which means that the CCR output model without explicit inputs has the infinite optimal objective function.

The BCC output oriented model without explicit inputs as follows:

$$\text{Max } \varphi$$

$$s.t. \sum_{j=1}^{n} \lambda_j \mathbf{Y_j} \geq \varphi \mathbf{Y_p}$$

$$\sum_{j=1}^{n} \lambda_j = 1$$

$$\lambda_j \geq 0, \, j = 1, \ldots, n \tag{2.40}$$

With a similar argument to that presented in model (2.38), model (2.40) can also be interpreted as an output-oriented CCR model in which all DMUs have an input equal to 1.

Note that if the DMU has no input (output) then CCR models cannot be used to calculate the relative efficiency of the units if no input (output) is provided for the DMUs. But if a fixed input (output), for example 1, is considered, then the CCR model can be used to calculate the efficiency of DMUs. While BCC models can be used for DMUs that have no input (output), it should be noted that this is equivalent to using CCR models that have a fixed value for all inputs (outputs). In addition to radial models, non-radial models can also be used to calculate relative efficiency.

2.5 Weight Restrictions in DEA

In basic DEA models, the weights are at the distance $(0, +\infty)$, and the models can select a value from this range. If the unit under evaluation is in a good position in one index compared to the others, the model chooses a very high weight for this index and a very low weight for other indicators in order to place the performance status of the unit under evaluation at its highest level. Excessive difference between the two corresponding weights of the two indicators is one of the disadvantages of these models that can show the evaluation results far from reality.

In general, there are the following major drawbacks to calculating weights in basic DEA models [23]:

a. Selecting a weight with a value of zero or ε.
b. Selecting a weight with a very high value.
c. There is a huge difference in calculating the two weights.
d. Incorrect order in the calculation between weights (contrary to the decision of the decision maker).

Accordingly, it is necessary to restriction the weights. The types of weight restrictions that can be added to basic DEA models are as follows:

$$\begin{cases} \mathbf{AV} + \mathbf{BU} \leq d \\ \alpha'_{il} \leq \frac{v_i}{v_l} \leq \alpha''_{il}, \ i, l = 1, \ldots, m \ \& \ \beta'_{rt} \leq \frac{u_r}{u_t} \leq \beta''_{rt}, \ r, t = 1, \ldots, s \end{cases} \tag{2.41}$$

If in the first constraint (2.41), $d = 0$, then the constraint is homogeneous and otherwise it is called inhomogeneous. The types of the second constraints are inhomogeneous. Obviously, the values of matrices $\mathbf{A}$ and $\mathbf{B}$ and scalars α', α'', β', β'' are determined by the relative importance between the indicators and by decision maker. Adding these constraints to multiplicative linear models can create problems, some of which are outlined below [24].

A. Infeasiblity
B. Failure to calculate relative efficiency
C. Calculation negative efficiency.

Note that adding any weight constraints is equivalent to adding a virtual unit to the observation set and creating a new PPS. So if this new virtual unit is fall in the previous PPS then the new and previous PPS will be the same. If the new virtual unit is not fall in the previous PPS then the new PPS will be larger than the old PPS. Therefore, PPS will not be smaller with the addition of a new virtual unit, and therefore the calculated relative efficiency will not be larger than the old calculated relative efficiency with the addition of the weight constraints.

The problems are often due to the fact that the weights of the DEA multiplicative linear programming model are derived from the variable change of fractional weights. Hence the addition of weight restrictions to the following model is the main goal.

$$\text{Max } RE_p = \frac{\frac{\mathbf{UY_p}}{\mathbf{VX_p}}}{\max\limits_{1 \leq j \leq n} \left\{ \frac{\mathbf{UY_j}}{\mathbf{VX_j}} \right\}}$$

$$s.t.\, \mathbf{U} \in A$$

$$\mathbf{V} \in B \tag{2.42}$$

If the weight restrictions are homogeneous, then adding these constraints to the relative efficiency model is equivalent to adding these constraints to the linear model. Therefore the general linear DEA models with the homogeneous weight restrictions is as follows:

$$\text{Max } \ \mathbf{UY_p} + u_0$$
$$s.t. \ \mathbf{VX_p} = 1$$
$$\mathbf{UY_j} + u_0 - \mathbf{VX_j} \leq 0, \ j = 1, \ldots, n$$
$$\mathbf{AU} + \mathbf{BV} \leq 0$$
$$u_0 \in F$$

$$(\mathbf{U}, \mathbf{V}) \geq \mathbf{1}\varepsilon \tag{2.43}$$

If inhomogeneous constraints are added to the DEA models then the following model will be obtained.

$$
\begin{aligned}
&\text{Max } \mathbf{U}\mathbf{Y_p} + u_0 \\
&s.t.\ \mathbf{V}\mathbf{X_p} = 1 \\
&\qquad \mathbf{U}\mathbf{Y_j} + u_0 - \mathbf{V}\mathbf{X_j} \leq 0,\ j = 1, \dots, n \\
&\qquad \mathbf{A}\mathbf{U} + \mathbf{B}\mathbf{V} = qd, \\
&\qquad q \geq \varepsilon \\
&\qquad u_0 \in F \\
&\qquad (\mathbf{U}, \mathbf{V}) \geq \mathbf{1}\varepsilon.
\end{aligned}
\tag{2.44}
$$

It should be noted that in model (2.44) there is the infeasibility and non-calculation of relative efficiency.

2.6   Summary

This chapter is dedicated to providing an introduction to DEA as a tool in the study of preferential voting. PPS and its potential principles were examined and with the acceptance or non-acceptance of some principles, traditional DEA models such as CCR, BCC, FDH and CHD were presented. The concept of relative efficiency was also expressed and dual of traditional models were presented based on this concept. Models without explicit inputs and outputs were expressed and related concepts were presented. Finally, the concept of weight restriction was expressed in the DEA and its variants.

References

1. Sharafi, H., Soltanifar, M., Lotfi, F.H.: Selecting a green supplier utilizing the new fuzzy voting model and the fuzzy combinative distance-based assessment method. EURO J. Decis. Process. **10**, 100010 (2022)
2. Soltanifar, M., Ghiyasi, M., Sharafi, H.: Inverse DEA-R models for merger analysis with negative data. IMA J. Manag. Math. **00**, 1–20 (2022)
3. Soltanifar, M., Sharafi, H.A.: Modified DEA cross efficiency method with negative data and its application in supplier selection. J. Comb. Optim. **43**, 265–296 (2022)
4. Soltanifar, M., Shahghobadi, S.: Classifying inputs and outputs in data envelopment analysis based on TOPSIS method and a voting model. Int. J. Bus. Analytics **1**(2), 48–63 (2014)
5. Soltanifar, M., Shahghobadi, S.: Survey on rank preservation and rank reversal in data envelopment analysis. Knowl.-Based Syst. **60**, 10–19 (2014)
6. Soltanifar, M., Farhadi, F.: An application of data envelopment analysis for measuring the relative efficiency in banking industry. Manag. Sci. Lett. **4**(5), 1021–1026 (2014)

7. HosseinzadehLotfi, F., Jahanshahloo, G.R., Soltanifar, M., Ebrahimnejad, A., Mansourzadeh, S.M.: Relationship between MOLP and DEA based on output-orientated CCR dual model. Expert Syst. Appl. **37**(6), 4331–4336 (2010)

8. Lotfi, F.H., Jahanshahloo, G.R., Ebrahimnejad, A., Soltanifar, M., Mansourzadeh, S.M.: Target setting in the general combined-oriented CCR model using an interactive MOLP method. J. Comput. Appl. Math. **234**(1), 1–9 (2010)

9. Ghiyasi, M., Soltanifar, M., Sharafi, H.: A novel inverse DEA-R model with application in hospital efficiency. Socio-Econ Plann. Sci. 101427 (2022)

10. Soltanifar, M., Lotfi, F.H., Sharafi, H., Lozano, S.: Resource allocation and target setting: a CSW–DEA based approach. Ann. Oper. Res. 2022.

11. Boles, J.: Efficiecny squared-efficient computation of efficiency indexes. Proc. Thirty Ninth Ann. Meet. West. Farm Econ. Assoc. **39**, 137–142 (1966)

12. Farrell, M.J.: The measurement of productive efficiency. J. Royal Stat. Soc.: Ser. A (General) **120**(3), 253–281 (1957)

13. Cooper, W.W., Seiford, L.M., Tone, K.: Data envelopment analysis: a comprehensive text with models, applications, references and DEA-solver software, vol. 2, p. 489. Springer, New York (2007)

14. Charnes, A., Cooper, W.W., Rhodes, E.: Measuring the efficiency of decision-making units. Eur. J. Oper. Res. **3**(4), 339–338 (1979)

15. Banker, R.D., Charnes, A., Cooper, W.W.: Some models for estimating technical and scale inefficiencies in data envelopment analysis. Manage. Sci. **30**(9), 1078–1092 (1984)

16. Tulkens, H.: On FDH efficiency analysis: some methodological issues and applications to retail banking, courts and Urban transit. In: Public Goods, Environmental Externalities and Fiscal Competition, pp. 311–342. Springer, Boston (2006)

17. Soltanifar, M., Jahanshahloo, G.R., Lotfi, F.H., Mansourzadeh, S.M.: On efficiency in convex hull of DMUs. Appl. Math. Model. **37**(4), 2267–2278 (2013)

18. Tone, K.: A slacks-based measure of efficiency in data envelopment analysis. Eur. J. Oper. Res. **130**(3), 498–509 (2001)

19. Thompson, R.G., Dharmapala, P.S., Thrall, R.M.: Linked-cone DEA profit ratios and technical efficiency with application to Illinois coal mines. Int. J. Prod. Econ. **39**(1–2), 99–115 (1995)

20. Charnes, A., Cooper, W.W.: Programming with linear fractional functionals. Naval Res. Logistics Q. **9**(3–4), 181–186 (1962)

21. Liu, W.B., Zhang, D.Q., Meng, W., Li, X.X., Xu, F.: A study of DEA models without explicit inputs. Omega **39**(5), 472–480 (2011)

22. Masoumzadeh, A., Toloo, M., Amirteimoori, A.: Performance assessment in production systems without explicit inputs: an application to basketball players. IMA J. Manag. Math. **27**(2), 143–156 (2016)

23. Podinovski, V.V.: Production trade-offs and weight restrictions in data envelopment analysis. J. Oper. Res. Soc. **55**(12), 1311–1322 (2004)

24. Razipour-GhalehJough, S., Lotfi, F.H., Jahanshahloo, G., Rostamy-Malkhalifeh, M., Sharafi, H.: Finding closest target for bank branches in the presence of weight restrictions using data envelopment analysis. Ann. Oper. Res. **288**(2), 755–787 (2020)

Chapter 3
Introduction to Fuzzy Logic

Abstract In real world, we sometimes face situations where we don't know which decision is right or wrong, and the correct action is hidden from view. At this time, "fuzzy logic" offers a flexible and valuable proposition. In this way, the amount of uncertainty can be determined for each situation. For this reason, fuzzy logic is sometimes called doubtful logic because its results are created with doubts. In this chapter, this logic begins with the presentation of fuzzy sets and continues with the definitions of fuzzy numbers. Finally, one of the applications of fuzzy numbers, i.e. converting linguistic terms into fuzzy numbers for decision making in fuzzy logic, is presented.

3.1 Introduction

Fuzzy logic is a form of multi-valued logic in which the logical value of variables can be any real number between 0 and 1. This logic is used to implement the concept of partial correctness, so that the degree of correctness can be any value between completely true and completely false. Fuzzy theory was introduced by Zadeh [1]. After a lot of expansion and deepening, this theory has had various applications in different fields. Many decision-making methods were rewritten with this logic and made it possible to reach more justified results. Many of these can be found in Kahraman et al. [2]. Also, a library review of decision making methods with probabilistic information by Liao et al. [3] presented that can be used in the study of the application of fuzzy theory in decision making. Recently, the theory of fuzzy sets has become a tool to deal with the problem of decision making in the presence of uncertainty. Reference [4] were among the pioneers who proposed the use of fuzzy data in the objective function and constraints of the optimization problem. Especially, there are different approaches to use DEA in fuzzy environments [5–11]. The application of this logic in software science can be simply defined as follows: Fuzzy logic goes beyond the logic of "0 and 1" values of classical software and opens a new door for the world of software science and computers [12]. Fuzzy logic extracts and applies new values "we may go" or "we go if …" or even "we will

© The Author(s), under exclusive license to Springer Nature Switzerland AG 2023   31
M. Soltanifar et al., *Preferential Voting and Applications: Approaches Based on Data Envelopment Analysis*, Studies in Systems, Decision and Control 471,
https://doi.org/10.1007/978-3-031-30403-3_3

probably go" from the space between two values "we go" or "we don't go". In this way, for example, the bank manager can go beyond the logic of "we will give a loan" or "we will not give a loan" and say: "we will give a loan if…" or "we will not give a loan but …" Such logic can justify the results of decision-making methods and turn them into a more powerful tool for decision support. In this chapter, definitions and preliminary principles of fuzzy logic are briefly presented.

3.1.1 Fuzzy Sets

Understanding the concept of "fuzzy set" is at the center of learning the application of fuzzy logic in management. In other words, fuzzy logic can be expressed as reasoning with fuzzy sets. Fuzzy sets can be described in comparison with classical sets and the concept of degree of membership. In a classical set, each element of the set has a membership degree of 1 and other elements have a membership degree of 0. It means that an element is either a member of set A or it is not a member. Unlike classical sets, fuzzy sets do not have clear and well-defined boundaries. For example, based on the concept of classical sets, a student who scores 9.5 (out of 20) is placed in the failed set, and a student who scores 10 is placed in the pass set. Now this question is raised: Is the difference between these two people to such an extent that the first person should lag behind the second person in the same way? This definition of an absolute threshold value for the inclusion of people in a set is in classical logic. Fuzzy logic has introduced new essential concepts in the form of uncertainty and ambiguity into mathematics and logic and has revolutionized the foundations of classical science. The key to understanding the fuzzy set is the concept of membership degree. Below are some basic definitions in this regard.

Definition 3.1 Let X be a reference set, the members of the set are defined by a property. Let P be a certain property. In this case, the subset A of X is displayed as Eq. (3.1).

$$A = \{x \in X | P(x)\} \tag{3.1}$$

For each x of X, i.e. x has property P, it can be either true or false. If true, the value 1 is taken and if false, the value 0, then the set A can be represented as Eq. (3.2).

$$A = \{(x, \mu_A(x)) | x \in X\}$$
$$\mu_A(x) : X \rightarrow \{0, 1\}$$
$$\mu_A(x) = \begin{cases} 1, & p(x) \text{ is true} \\ 0, & p(x) \text{ is false} \end{cases} \tag{3.2}$$

Definition 3.2 A fuzzy subset $\tilde{A}$ of the reference set X is defined as Eq. (3.3), where $\mu_{\tilde{A}}$ is the membership degree of the set.

$$A = \{(x, \mu_{\tilde{A}}(x)) | x \in X\}$$
$$\mu_{\tilde{A}}(x) : X \rightarrow [0, 1] \tag{3.3}$$

Definition 3.3 Suppose X is the reference set and $\tilde{A}$ is a fuzzy subset of X with membership function $\mu_{\tilde{A}}$, in this case the support set of $\tilde{A}$ is defined as (3.4).

$$Supp\left(\tilde{A}\right) = \{x \in X | \mu_{\tilde{A}}(x) > 0\} \tag{3.4}$$

Definition 3.4 Suppose X is the reference set and $\tilde{A}$ is a fuzzy subset of X with membership function $\mu_{\tilde{A}}$, the height of $\tilde{A}$ is defined as (3.5).

$$M = Sup\{\mu_{\tilde{A}}(x) | x \in X\} \tag{3.5}$$

Definition 3.5 A fuzzy set is normal when among its members there is at least one member with membership degree of one. In other words, its height should be one. In fact, proposition (3.6) must be true.

$$\exists x \ (x \in X \& \mu_{\tilde{A}}(x) = 1) \tag{3.6}$$

To normalize a non-normal set, the membership degree of all elements can be divided by the largest membership degree (height) of the set.

Definition 3.6 The cardinality of set $\tilde{A}$ is defined as Eq. (3.7).

$$\left|\tilde{A}\right| = \begin{cases} \sum_{x \in X} \mu_{\tilde{A}}(x), & X \text{ is a finite set} \\ \int_X \mu_{\tilde{A}}(x), & X \text{ is an infinite set} \end{cases} \tag{3.7}$$

Definition 3.7 The relativity number of set $\tilde{A}$ is defined as Eq. (3.8).

$$\|A\| = \frac{\left|\tilde{A}\right|}{|X|} \tag{3.8}$$

Example **3.1** Assume $X = \{1, 2, ..., 10\}$ and $\tilde{A} = \{(1, 0.5), (2, 0.7), (3, 1), (4, 0.1), (5, 0.2), (9, 0.2), (10, 0.7)\}$. In this case, the following can be provided:

$$Supp(\tilde{A}) = \{1, 2, 3, 4, 5, 9, 10\}$$
$$M = Sup\{0.5, 0.7, 0.1, 1, 0.2, 0.2, 0.7\} = 1$$
$$\left|\tilde{A}\right| = 0.5 + 0.7 + 0.1 + 1 + 0.2 + 0.2 + 0.7 = 3.4$$

$$\left\|\tilde{A}\right\| = \frac{\left|\tilde{A}\right|}{|X|} = \frac{3.4}{10} = 0.34$$

Definition 3.7 $\alpha-$cut of a fuzzy set contains elements that have membership degree at least equal to α. . This concept can be seen in (3.9).

$$\tilde{A}_\alpha = \left\{x \in X | \mu_{\tilde{A}}(x) \geq \alpha\right\} \tag{3.9}$$

Definition 3.8 Strong $\alpha -$ cut of a fuzzy set is defined as Eq. (3.10).

$$\tilde{A}_{\overline{\alpha}} = \left\{x \in X | \mu_{\tilde{A}}(x) > \alpha\right\} \tag{3.10}$$

Example 3.2 Considering example (3.1), the following can be written:

$$\tilde{A}_{0.7} = \{2, 3, 10\}$$
$$\tilde{A}_{\overline{0.7}} = \{3\}$$

Definition 3.9 Two fuzzy sets $\tilde{A}$ and $\tilde{B}$ are called equal if and only if statement (3.11) is true.

$$\forall x \quad \left(x \in X \Rightarrow \mu_{\tilde{A}}(x) = \mu_{\tilde{B}}(x)\right) \tag{3.11}$$

Definition 3.10 Fuzzy set $\tilde{B}$ is called a subset of fuzzy set $\tilde{A}$ if and only if statement (3.12) is true.

$$\forall x \quad \left(x \in X \Rightarrow \mu_{\tilde{A}}(x) \leq \mu_{\tilde{B}}(x)\right) \tag{3.12}$$

Definition 3.11 The set $\tilde{A}$ is called convex if and only if statement (3.13) is true.

$$\forall x_1 \forall x_2 \forall \lambda \left(x_1, x_2 \in X \& \lambda \in [0, 1], \quad \mu_{\tilde{A}}[\lambda x_1 + (1 - \lambda)x_2] \quad \geq min\left\{\mu_{\tilde{A}}(x_1), \mu_{\tilde{A}}(x_2)\right\}\right) \tag{3.13}$$

In fact, $\tilde{A}$ is convex if and only if $\tilde{A}_\alpha$ is convex for every $\alpha \in [0, 1]$.

Definition 3.12 Suppose $\tilde{A}$ and $\tilde{B}$ are two fuzzy sets with membership functions equal to $\mu_{\tilde{A}}$ and $\mu_{\tilde{B}}$, respectively, in this case the operators of union, Intersection, complement and Cartesian product are defined as Eqs. (3.14).

$$\forall x \left(x \in X \Rightarrow \mu_{\tilde{A} \cap \tilde{B}}(x) = min\left\{\mu_{\tilde{A}}(x), \mu_{\tilde{B}}(x)\right\}\right)$$
$$\forall x \left(x \in X \Rightarrow \mu_{\tilde{A} \cup \tilde{B}}(x) = max\left\{\mu_{\tilde{A}}(x), \mu_{\tilde{B}}(x)\right\}\right)$$
$$\forall x \left(x \in X \Rightarrow \mu_{\overline{\tilde{A}}}(x) = 1 - \mu_{\tilde{A}}(x)\right)$$
$$\forall x \left(x \in X \Rightarrow \mu_{\tilde{A} \times \tilde{B}}(a, b) = min\left\{\mu_{\tilde{A}}(a), \mu_{\tilde{B}}(b)\right\}\right) \tag{3.14}$$

Example 3.3 Consider a city that has 3 universities $X = \{a, b, c\}$. If fuzzy set $\tilde{A}$ are universities with high scientific level and fuzzy set $\tilde{B}$ are universities with high geographical location and are presented as $\tilde{A} = \{(a, 0.4), (b, 0.6), (c, 0.7)\}$ and $\tilde{B} = \{(a, 0.5), (c, 0.8)\}$, then the effect of operators (3.14) on these two sets are as follows:

$$\tilde{A} \cup \tilde{B} = \{(a, 0.5), (b, 0.6), (c, 0.8)\}$$
$$\tilde{A} \cap \tilde{B} = \{(a, 0.4), (c, 0.7)\}$$
$$\bar{\tilde{A}} = \{(a, 0.6), (b, 0.4), (c, 0.3)\} \& \bar{\tilde{B}} = \{(a, 0.5), (b, 1), (c, 0.2)\}$$
$$\tilde{A} \times \tilde{B} = \{((a, a), 0.4), ((a, c), 0.4), ((b, a), 0.5), ((b, c), 0.6),$$
$$((c, a), 0.5), ((c, c), 0.7)\}$$

Definition (3.12) can be extended to the case of more than two sets. For example, if $X_1, X_2, \ldots, X_n$ are reference sets and $X = X_1 \times X_2 \times \ldots \times X_n$ is their Cartesian product, and also if $\tilde{A}_1, \tilde{A}_2, \ldots, \tilde{A}_n$ are fuzzy subsets of $X_1, X_2, \ldots, X_n$, then the Cartesian product of these n sets is defined in such a way that its membership function is as follows:

$$\mu_{\tilde{A}_1 \times \tilde{A}_2 \times \ldots \times \tilde{A}_n}(a_1, a_2, \ldots, a_n) = \min\{\mu_{\tilde{A}_1}(a_1), \mu_{\tilde{A}_2}(a_2), \ldots, \mu_{\tilde{A}_n}(a_n)\}$$

Definition 3.13 (*The extension principle*): Suppose that $X_1, X_2, \ldots, X_n$ are reference sets and $X = X_1 \times X_2 \times \ldots \times X_n$ is their Cartesian product, and also suppose $\tilde{A}_1, \tilde{A}_2, \ldots, \tilde{A}_n$ are fuzzy subsets of $X_1, X_2, \ldots, X_n$ and $f : X \to Y$ is a function from X to Y. Then, the result of the operation of f on n of the fuzzy sets $\tilde{A}_1, \tilde{A}_2, \ldots, \tilde{A}_n$ is defined as follows, which is the subset of the fuzzy set $\tilde{B}$ of Y.

$$\tilde{B} = f\left(\tilde{A}_1, \ldots, \tilde{A}_n\right)$$
$$= \{(y, \mu_{\tilde{B}}(y)) | y = f(x_1, \ldots, x_n)$$
$$\& x_1 \in X_1 \& \ldots \& x_n \in X_n\}$$

$$\text{where } \mu_{\tilde{B}}(y) = \begin{cases} \underset{\substack{(x_1, \ldots, x_n) \\ y = f(x_1, \ldots, x_n)}}{Sup} \left(\min\{\mu_{\tilde{A}_1}(x_1), \ldots, \mu_{\tilde{A}_n}(x_n)\}\right), & f^{-1}(y) \neq \emptyset \\ 0, & f^{-1}(y) = \emptyset \end{cases}$$

Example 3.4 In example (3.3), suppose $f : X \to X$ is given as $f(x_1, x_2) = x_1 + x_2$. In this case, according to the extension principle, $f\left(\tilde{A}, \tilde{B}\right)$ will be as follows:

$$f\left(\tilde{A}, \tilde{B}\right) = \{((a + a), 0.4), ((b + a), 0.5), ((b + c), 0.6),$$
$$((c + a), 0.5), ((c + c), 0.7)\}, (a + a \neq b + c)$$

3.2 Fuzzy Numbers

Fuzzy numbers are generalizations of crisp numbers for which algebraic operators can be generalized by the extension principle for them.

Definition 3.14 A fuzzy set like $\tilde{N}$ from $\mathbb{R}$ is called a fuzzy number if and only if

i. $\tilde{N}$ is convex.
ii. $\mu_{\tilde{N}}$ is single-dimensional. (That is, only $x^o \in \mathbb{R}$ exists so that $\mu_{\tilde{N}}(x^o) = 1$).
iii. $\mu_{\tilde{N}}$ is piecewise continuous.

Example 3.5 Consider the fuzzy sets $\tilde{A} = \{(3, 0.2), (4, 0.6), (5, 1), (6, 0.7), (7, 0.1)\}$, $\tilde{B} = \{(8, 0.3), (9, 0.7), (10, 0.8), (11, 0.7), (12, 0.9)\}$ and $\tilde{C} = \{(3, 0.8), (4, 1), (5, 1), (6, 0.7)\}$. The set $\tilde{A}$ is a fuzzy number because it has the three characteristics of definition (3.14). But the set $\tilde{B}$ is not a fuzzy number because none of the degrees of membership is equal to 1, and it is also not convex, and for example, the degree of membership of number 12 is greater than the degree of membership of number 11. Finally, the set $\tilde{C}$ is not a fuzzy number because the degree of membership of the numbers 4 and 5 is both equal to 1.

In the literature of fuzzy numbers, the concepts of triangular fuzzy number and trapezoidal fuzzy number are usually used more. In this chapter, the definitions and concepts presented are focused on triangular fuzzy numbers.

Definition 3.15 A triangular fuzzy number (TFN) is a number such as $\tilde{A}$ with membership function $\mu_{\tilde{A}}(x) : \mathbb{R} \to [0, 1]$ is in the form of Eq. (3.15) [13].

$$\mu_{\tilde{A}}(x) = \begin{cases} \frac{x-l}{m-l}, & l \leq x \leq m \\ \frac{u-x}{u-m}, & m \leq x \leq u \\ 0, & Otherwise \end{cases} \tag{3.15}$$

In definition (3.15), l, u, and m are the lower bound, the upper bound, and the core of the fuzzy number $\tilde{A}$, respectively. This fuzzy number has been demonstrated in Fig. 3.1. The TFN $\tilde{A}$ is illustrated in the form of $\tilde{A} \approx (l, m, u)$.

The TFNs are intuitive, easy to use, computationally simple, and useful in representing ambiguity in a fuzzy environment [14–16].

Definition 3.16 The defuzzification of TFN $\tilde{A} \approx (l, m, u)$ based on the center of area method is defined as follows [13]:

$$V\left(\tilde{A}\right) = l + \frac{(u - l) + (m - l)}{3} = \frac{l + m + u}{3} \tag{3.16}$$

Definition 3.17 The arithmetic operations between two TFNs are defined as follows [13]:

Let $\tilde{A}_1 \approx (l_1, m_1, u_1)$ and $\tilde{A}_2 \approx (l_2, m_2, u_2)$, be two TFNs and $k \in \mathbb{R}^{\geq 0}$ then.

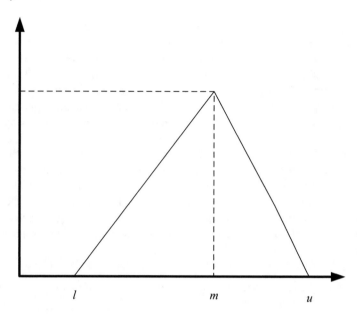

Fig. 3.1 Triangular fuzzy number membership function

$$k\tilde{A} \approx (kl, km, ku) \tag{3.17}$$

$$\tilde{A}_1 \oplus \tilde{A}_2 \approx (l_1 + l_2, m_1 + m_2, u_1 + u_2) \tag{3.18}$$

$$\tilde{A}_1 \ominus \tilde{A}_2 \approx (l_1 - u_2, m_1 - m_2, u_1 - l_2) \tag{3.19}$$

Definition 3.18 A TFN $\tilde{A} \approx (l, m, u)$ is known as positive when $l > 0$ [13].

Definition 3.19 The inverse of positive TFN $\tilde{A} \approx (l, m, u)$, is defined as follows [13]:

$$\tilde{A}^{-1} \approx \left(\frac{1}{u}, \frac{1}{m}, \frac{1}{l}\right) \tag{3.20}$$

Definition 3.20 For the two positive TFNs $\tilde{A}_1 \approx (l_1, m_1, u_1)$ and $\tilde{A}_2 \approx (l_2, m_2, u_2)$, the multiplication and division of the two TFNs are defined in the form of Eqs. (3.21) and (3.22) [13].

$$\tilde{A}_1 \otimes \tilde{A}_2 \approx (l_1 \times l_2, m_1 \times m_2, u_1 \times u_2) \tag{3.21}$$

$$\frac{\tilde{A}_1}{\tilde{A}_2} \approx \tilde{A}_1 \otimes \tilde{A}_2^{-1} \approx \left(\frac{l_1}{u_2}, \frac{m_1}{m_2}, \frac{u_1}{l_2}\right) \tag{3.22}$$

Ranking fuzzy numbers is a very important decision method in decision analysis based on the logic of uncertainty. Recently, a number of approaches to ranking fuzzy numbers have been investigated, some of which are non-intuitive and inconsistent. The first method for ranking fuzzy numbers was proposed by Jain [17]. Reference [18] then proposed an approach for ranking fuzzy numbers using maximizing sets and minimizing set concepts. Reference [19] developed a ranking approach based on an integrated value index to overcome the shortcomings of Chen's approach. Reference [20] reviewed the approach proposed by Liou and Wang [19], based on the convex combination of right and left integral values through an index of optimism, enumerated some related shortcomings, and then developed a modified one ranking approach. The median value ranking approach introduces new values of left, right, and total integral values of the fuzzy numbers and is further applied to differentiate fuzzy numbers that have the compensation of areas. Other ranking methods can also be found in [21–27]. In this chapter, the proposed ranking [20] is presented, which is as follows:

In assuming that, $\tilde{A}_i \approx (l_i, m_i, u_i)$, $i = 1, 2, \ldots, n$, are n TFNs and $x_{min} = \min\limits_{1 \le i \le n} l_i$. In which case, the left, right, and the total integrals will be defined in the form of Eqs. (3.23), (3.24) and (3.25).

$$S_L\left(\tilde{A}_i\right) = (m_i - x_{\min}) - \int_{l_i}^{m_i} \frac{x - l_i}{m_i - l_i} dx$$
$$= \frac{1}{2}(l_i + m_i - 2x_{\min}); i = 1, 2, \ldots, n \qquad (3.23)$$

$$S_R\left(\tilde{A}_i\right) = (m_i - x_{\min}) + \int_{m_i}^{u_i} \frac{u_i - x}{u_i - m_i} dx$$
$$= \frac{1}{2}(m_i + u_i - 2x_{\min}); i = 1, 2, \ldots, n \qquad (3.24)$$

$$S_T\left(\tilde{A}_i\right) = \frac{1}{2}\left(S_L\left(\tilde{A}_i\right) + S_R\left(\tilde{A}_i\right)\right)$$
$$= \frac{1}{4}(l_i + 2m_i + u_i - 4x_{\min}); i = 1, 2, \ldots, n \qquad (3.25)$$

The proposed approach uses $S_T\left(\tilde{A}_i\right)$ to rank fuzzy numbers. The larger $S_T\left(\tilde{A}_i\right)$ is, the larger is the TFN $\tilde{A}_i$. Therefore, for any TFNs $\tilde{A}_i$ and $\tilde{A}_j$, we have the following properties.

1. If, $S_T\left(\tilde{A}_i\right) > S_T\left(\tilde{A}_j\right)$, then, $\tilde{A}_i \succ \tilde{A}_j$.
2. If, $S_T\left(\tilde{A}_i\right) < S_T\left(\tilde{A}_j\right)$, then, $\tilde{A}_i \prec \tilde{A}_j$.
3. If, $S_T\left(\tilde{A}_i\right) = S_T\left(\tilde{A}_j\right)$, then, $\tilde{A}_i \approx \tilde{A}_j$.

Example **3.6** Applying the above definitions to TFNs $\tilde{A}_1 = (-1, 0, 1)$, $\tilde{A}_2 = (1, 2, 3)$ and $\tilde{A}_3 = (2, 3, 4)$ gives the following results:

$$V\left(\tilde{A}_1\right) = \frac{-1 + 0 + 1}{3} = 0; \ V\left(\tilde{A}_2\right) = \frac{1 + 2 + 3}{3} = 2;$$
$$V\left(\tilde{A}_3\right) = \frac{2 + 3 + 4}{3} = 4.5$$

$$2\tilde{A}_1 \approx (-2, 0, 2); \ 2\tilde{A}_2 \approx (2, 4, 6); \ 2\tilde{A}_3 \approx (4, 6, 8)$$

$$\tilde{A}_1 \oplus \tilde{A}_2 \approx (0, 2, 4); \ \tilde{A}_1 \oplus \tilde{A}_3 \approx (1, 3, 5); \ \tilde{A}_2 \oplus \tilde{A}_3 \approx (3, 5, 7)$$

$$\tilde{A}_1 \ominus \tilde{A}_2 \approx (-4, -2, 0); \ \tilde{A}_1 \ominus \tilde{A}_3 \approx (-5, -3, -1); \ \tilde{A}_2 \ominus \tilde{A}_3 \approx (-3, -1, 1)$$

$$\tilde{A}_1 \prec 0; \ \tilde{A}_2 \succ 0; \ \tilde{A}_3 \succ 0$$

$$\tilde{A}_2^{-1} \approx \left(\frac{1}{3}, \frac{1}{2}, 1\right); \ \tilde{A}_3^{-1} \approx \left(\frac{1}{4}, \frac{1}{3}, \frac{1}{2}\right)$$

$$\tilde{A}_2 \otimes \tilde{A}_3 \approx (2, 6, 12)$$

$$\frac{\tilde{A}_2}{\tilde{A}_3} \approx \tilde{A}_2 \otimes \tilde{A}_3^{-1} \approx \left(\frac{1}{4}, \frac{2}{3}, \frac{3}{2}\right)$$

$$x_{min} = \min_{1 \leq i \leq 3} l_i = -1$$

$$S_L\left(\tilde{A}_1\right) = \frac{1}{2}(-1 + 0 - 2 \times (-1)) = \frac{1}{2}; S_L\left(\tilde{A}_2\right)$$
$$= \frac{1}{2}(1 + 2 - 2 \times (-1)) = \frac{5}{2}; \ S_L\left(\tilde{A}_3\right)$$
$$= \frac{1}{2}(2 + 3 - 2 \times (-1)) = \frac{7}{2}$$

$$S_R\left(\tilde{A}_1\right) = \frac{1}{2}(0 + 1 - 2 \times (-1)) = \frac{3}{2}; \ S_R\left(\tilde{A}_2\right)$$
$$= \frac{1}{2}(2 + 3 - 2 \times (-1)) = \frac{7}{2}; \ S_R\left(\tilde{A}_3\right)$$
$$= \frac{1}{2}(3 + 4 - 2 \times (-1)) = \frac{9}{2}$$

$$S_T\left(\tilde{A}_1\right) = \frac{1}{2}\left(\frac{1}{2} + \frac{3}{2}\right) = 1; \ S_T\left(\tilde{A}_2\right) = \frac{1}{2}\left(\frac{5}{2} + \frac{7}{2}\right) = 3;$$

$$S_T\left(\tilde{A}_3\right) = \frac{1}{2}\left(\frac{7}{2} + \frac{9}{2}\right) = 4$$

$$\tilde{A}_1 \prec \tilde{A}_2 \prec \tilde{A}_3$$

3.3   Converting Linguistic Terms to Fuzzy Numbers

In fuzzy theory, suitable conditions are provided to act in the conditions of uncertainty and imprecise systems, including the provision of conditions for the use of different linguistic terms that are used for variables with imprecise and ambiguous values. For example, in any language, it is less common to say "smokers are two and a half times more likely to get lung cancer"; rather, it is said that "smokers are more prone to lung cancer than other people". Linguistic terms such as "more" and the like express imprecise concepts that increase the citation of the expressed sentence. In short, linguistic terms mean words such as high, low, average, less, more, etc., which are used to express vague and imprecise concepts. There are different scales to convert linguistic terms into fuzzy numbers. Reference [28] presented various scales for this purpose, which are shown in Figs. 3.2, 3.3, 3.4, 3.5, 3.6, 3.7 and 3.8. Of course, these are not the only scales for converting linguistic terms into fuzzy numbers, and other scales can be provided if necessary. It should be noted that even when the number of words in two different variables is the same, the actual linguistic terms may be slightly different. For example, when we use similar words such as "high", their fuzzy numbers will be different from one scale to another, and this is because a linguistic term has different meanings in different situations. Many researchers used the concept of linguistic terms and fuzzy logic in the decision making process; because this concept uses qualitative words in fuzzy logic and will provide more substantiated results to support the decision [29].

3.4   Summary

In this chapter, an introduction to the fuzzy theory, which is a theory for action under uncertainty, was presented. This theory is able to mathematically transform many concepts, variables and systems that are imprecise and ambiguous, as is the case in most cases in the real world, and provide the basis for reasoning, inference, control and decision-making in conditions of uncertainty. In this chapter, basic definitions about fuzzy sets and fuzzy numbers were presented, and fuzzy operators based on triangular fuzzy numbers were presented. The concept of linguistic terms and how to convert them into fuzzy numbers was also explained.

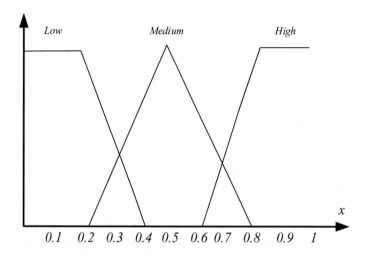

Fig. 3.2 Linguistic terms 1

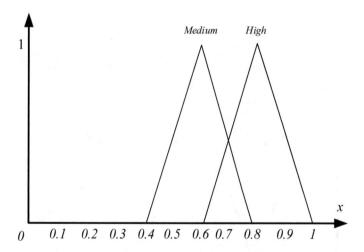

Fig. 3.3 Linguistic terms 2

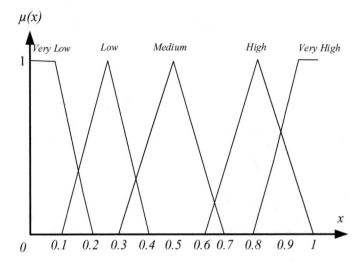

Fig. 3.4 Linguistic terms 3

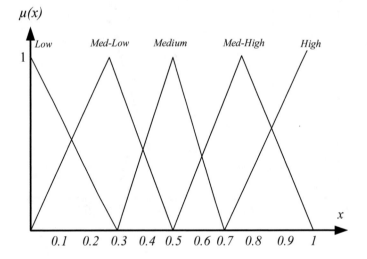

Fig. 3.5 Linguistic terms 4

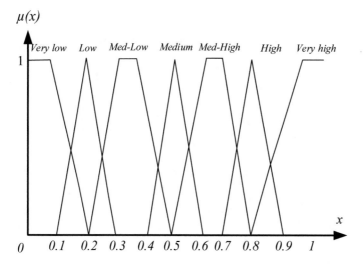

Fig. 3.6 Linguistic terms 5

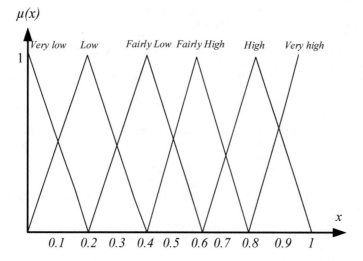

Fig. 3.7 Linguistic terms 6

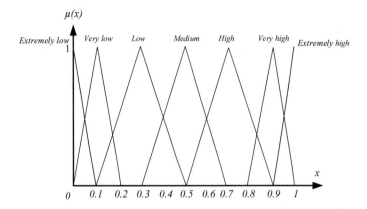

Fig. 3.8 Linguistic terms 7

References

1. Zadeh, L.A.: Fuzzy sets. Inf. Control **8**(3), 338–353 (1965)
2. Kahraman, C., Onar, S.C., Oztaysi, B.: Fuzzy multi criteria decision making: a literature review. Int. J. Comput. Intell. Syst. **8**(4), 637–666 (2015)
3. Liao, H., Mi, X., Xu, Z.: A survey of decision-making methods with probabilistic linguistic information: bibliometrics, preliminaries, methodologies, applications and future directions. Fuzzy Optim. Decis. Making **19**, 81–134 (2020)
4. Bellman, R.E., Zadeh, L.A.: Decision-making in a fuzzy environment. Manage. Sci. **17**(4), B141–B164 (1970)
5. Sengupta, J.K.: A fuzzy systems approach in data envelopment analysis. Comput. Math. Appl. **24**(8–9), 259–266 (1992)
6. Hougaard, J.L.: Fuzzy scores of technical efficiency. Eur. J. Oper. Res. **115**(3), 529–541 (1999)
7. Guh, Y.Y.: Data envelopment analysis in fuzzy environment. Int. J. Inf. Manage. Sci. **12**(2), 51–65 (2001)
8. Lertworasirikul, S., Fang, S.C., Joines, J.A., Nuttle, H.L.: Fuzzy data envelopment analysis (DEA): a possibility approach. Fuzzy Sets Syst. **139**(2), 379–394 (2003)
9. Wen, M., Li, H.: Fuzzy data envelopment analysis (DEA): model and ranking method. J. Comput. Appl. Math. **223**(2), 872–878 (2009)
10. Tavana, M., Shiraz, R.K., Hatami-Marbini, A., Agrell, P.J., Paryab, K.: Fuzzy stochastic data envelopment analysis with application to base realignment and closure (BRAC). Expert Syst. Appl. **39**(15), 12247–12259 (2012)
11. Tavana, M., Khalili-Damghani, K., Arteaga, F.J.S., Hossaini, A.: A fuzzy multi-objective multi-period network DEA model for efficiency measurement in oil refineries. Comput. Ind. Eng. **135**, 143–155 (2019)
12. Allahviranloo, T.: Uncertain information and linear systems. Springer Cham (2020)
13. Kaufmann, A., Gupta, M.M.: Introduction to fuzzy arithmetic theory and applications. Van Nostrand Reinhold, New York (1985)
14. Fung, R.Y.K., Chen, Y., Tang, J.: Estimating the functional relationships for quality function deployment under uncertainties. Fuzzy Sets Syst. **157**, 98–120 (2006)
15. Borovička, A.: New approach for estimation of criteria weights based on a linguistic evaluation. Expert Syst. Appl. **125**, 100–111 (2019)
16. Wang, F.: Preference degree of triangular fuzzy numbers and its application to multi-attribute group decision making. Expert Syst. Appl. **178**, 114982 (2021)

17. Jain, R.: Decision-making in the presence of fuzzy variables. IEEE Trans. Syst. Man Cybern. **6**, 698–703 (1976)
18. Chen, S.H.: Ranking fuzzy numbers with maximizing set and minimizing set. Fuzzy Sets Syst. **17**, 113–129 (1985)
19. Liou, T.S., Wang, M.J.: Ranking fuzzy numbers with integral value. Fuzzy Sets Syst. **50**, 247–255 (1992)
20. Vincent, F.Y., Dat, L.Q.: An improved ranking method for fuzzy numbers with integral values. Appl. Soft Comput. **14**, 603–608 (2014)
21. Khezerloo, S., Allahviranloo, T., Khezerloo, M.: Ranking of fuzzy numbers based on alpha-distance," In: 7th conference of the European Society for Fuzzy Logic and Technology (EUSFLAT-11). Aix-les-Bains (2011)
22. Allahviranloo, T., Abbasbandy, S., Saneifard, R.: An approximation approach for ranking fuzzy numbers based on weighted interval-value. Math. Comput. Appl. **16**, 588–597 (2011)
23. Allahviranloo, T., Jahantigh, M.A., Hajighasemi, S.: A new distance measure and ranking method for generalized trapezoidal fuzzy numbers. Math. Probl. Eng. Special Issue: Ranking Fuzzy Numbers and Its Extensions, 623757 (2013)
24. Ezadi, P.H., Allahviranloob, T.: New multi-layer method for z-number ranking using hyperbolic tangent function and convex combination. Intell. Autom. Soft Comput. **24**(1), 217–221 (2018)
25. Allahviranloo, T., Ezadi, S.: Z-Advanced numbers processes. Inf. Sci. **480**, 130–143 (2019)
26. Allahviranloo, T., Abbasbandy, S.S.R.: A method for ranking of fuzzy numbers using new weighted distance. Math. Comput. Appl. **16**(2), 359–369 (2011)
27. Allahviranloo, T., Firozja, M.A.: Ranking of fuzzy numbers by a new metric. Soft. Comput. **14**, 773–782 (2010)
28. Chen, S.J., Hwang, C.L., Fuzzy Multiple Attribute Decision Making. Springer (1992)
29. Sharafi, H., Soltanifar, M., HosseinzadehLotfi, F.: Selecting a green supplier utilizing the new fuzzy voting model and the fuzzy combinative distance-based assessment method. EURO J. Decis. Process. **10**, 100010 (2022)

Chapter 4
Preferential Voting Based on Data Envelopment Analysis

Abstract Voting is always considered as one of the best group decision making methods in decision science. How voters' votes are aggregated in the preferential voting process has a direct impact on the final voting results and is a very important issue. In this chapter, a suitable voting model is presented using the technique of Data Envelopment Analysis (DEA), which makes the voting results more justified and logical. The use of weight restrictions and the use of the concept of discrimination intensity functions increase the interaction with the decision maker in the presented model and turn the model into a powerful tool for decision support.

4.1 Introduction

Decision-making is one of the most important and fundamental tasks of management and the realization of organizational goals depends on its quality. So, from the point of view of experts in this field, decision-making is the main essence of management. What leads to the desirability of the results of a decision is the power of the decision support tool. A tool that supports the decision maker in choosing between two or more alternatives. The problem of "election" using the aggregation of voters' votes is one of the most important problems of group decision-making for which several models have been presented so far. Consider a group of people who need to make a group decision. Given that people have different opinions and tastes, how can they make a decision that includes all individual opinions? From such a simple issue to the election of politicians, they all need a group decision. Voting is a method of aggregating individual votes to reach a group decision. The structure of ballots are generally divided into two categories. In the first category, voters vote for one candidate, but in the second category, voters vote for more than one candidate. The second category is divided into two subcategories, one is that only the names of a few candidates are written on the ballot, and the second is that in addition to choosing a few candidates, the voters also express their preferences. In the mechanism of non-preference voting, voters choose r ($r < n$) people from among n candidates. So each voter votes for at most r candidates on the ballot, and the candidate who gets the most

© The Author(s), under exclusive license to Springer Nature Switzerland AG 2023
M. Soltanifar et al., *Preferential Voting and Applications: Approaches Based on Data Envelopment Analysis*, Studies in Systems, Decision and Control 471,
https://doi.org/10.1007/978-3-031-30403-3_4

votes is the winner. One of the disadvantages of this voting method, which is one of the most common voting methods, is that we have not considered the voting preferences and therefore the voters cannot convey their preferences to the society. In other words, a candidate may claim that despite receiving fewer votes, that candidate has been placed in the top voting priorities more than others and therefore has more merit for election. In classic voting models, votes are aggregated regardless of voting priority. Therefore, the result will not reflect the will of the voters. In preferential voting, more information is used from voters' opinions than in other electoral systems. In this type of voting, voters are not only asked to choose their favorite candidate, but they are also asked to nominate a second candidate if their first choice does not win, and the same question is asked: if the first choice And if their second one does not win, what candidate will be their third choice? Therefore, in the preferential voting system, each voter selects a subset of candidates and arranges them according to their preferences. In classic voting models, votes are aggregated regardless of voting priority, while in preferential voting it is these preferences that will play a major role in the final result. Several methods have been proposed for aggregating voters' votes in such a structure. Preferential voting is a mechanism where people's preferences can be used for a group decision. In preferential voting, the voter is asked to rank the candidates according to their preferences. This technique has many uses in decision-making issues. In this chapter, a linear programming based on the DEA policy is presented for ranking candidates in preference voting. Also, a sensitivity analysis on the model is presented with a numerical example.

4.2   Preferential Voting and Relative Efficiency

In the Chap. 1, there was a discussion about the famous methods of voting, as it was stated there that some methods are more privileged than the others. The methods based on Borda Count are more accepted among researchers than the rest of the methods and it can even be said that they created various applications of preferential voting as a tool in different prioritization. In this section, predetermined weights are not used to weight the voting priorities, but based on the DEA policy and with an optimistic view, these weights are determined by each candidate individually. In fact, as DEA was used to determine the relative efficiency, choosing the best weights to maximize the efficiency of each DMU will be the same in determining the final weight of each candidate.

To express this issue, it is assumed that after the preferential voting in which the voters have arranged n candidates in the first priority to the kth priority ($k \leq n$). The number of votes of candidate j ($j = 1, \ldots, n$) in the priority of rth ($r = 1, \ldots, k$) is displayed with v_{rj}. Therefore, the overall desirability index of candidate j is equal to $Z_j = \sum_{r=1}^{k} w_r v_{rj}$. In this regard, the weights $w_r (r = 1, \ldots, k)$ must be in order. There have been many interpretations of this arrangement of weights. In other words, relation (4.1) can be presented for the order of the weights.

$$w_1 \succ w_2 \succ w_3 \succ \ldots \succ w_k \tag{4.1}$$

"$\succ$" can be presented as "$\geq$" or "$>$" or other forms. "$\geq$" is called weak order and "$>$" is called strong order.

If the weights of the voting priorities are known, $\sum_{r=1}^{k} w_r v_{rp}$ can be a criterion for determining the final rank of the assumed candidate p. But the final rank is not supposed to be determined with predetermined weights. In order to present this criterion in the form of a relative criterion, relation (4.2) is used. Based on this, the value assigned to each candidate will be in the $[0, 1]$.

$$Z_p = \frac{\sum_{r=1}^{k} w_r v_{rp}}{\underset{1 \leq j \leq n}{\text{Max}} \sum_{r=1}^{k} w_r v_{rj}} \tag{4.2}$$

DEA policy is used to determine the weights of voting priorities and achieve an efficiency value for each candidate. This policy should be considered together with the order (4.1) for the weights. Thus, the model (4.3) is presented to determine the efficiency value of the assumed candidate p.

$$\text{Max} \quad Z_p = \frac{\sum_{r=1}^{k} w_r v_{rp}}{\underset{1 \leq j \leq n}{\text{Max}} \sum_{r=1}^{k} w_r v_{rj}}$$

$$s.t. \quad w_1 \succ w_2 \succ w_3 \succ \ldots \succ w_k \succ 0 \tag{4.3}$$

Model (4.3) can be converted to model (4.4) by applying Charnes and Cooper variable change $\left(\underset{1 \leq j \leq n}{\text{Max}} \sum_{r=1}^{k} w_r v_{rj} = 1 \right)$.

$$\text{Max} \quad Z_p = \sum_{r=1}^{k} w_r v_{rp}$$

$$s.t. \quad \sum_{r=1}^{k} w_r v_{rj} \leq 1, \quad j = 1, 2, \ldots n$$

$$w_1 \succ w_2 \succ w_3 \succ \ldots \succ w_k \succ 0 \tag{4.4}$$

In fact, the model (4.4) is a weight selection model for the pth candidate. Based on the optimistic policy of DEA and by observing the priority given for the weights, the highest total weight is selected for the candidate p, which is actually an efficiency value for that candidate because it is located in the interval $[0, 1]$. Cook and Kress [1] presented the model (4.4) using the optimistic DEA policy to select weights based on Borda Count. Inspired by Thompson et al. [2, 3] in choosing the assurance region, they proposed the structure (4.5) for the weight restriction of the model (4.4).

$$w_r - w_{r+1} \geq d(r, \varepsilon) \quad r = 1, \ldots, k - 1$$
$$w_k \geq d(k, \varepsilon) \tag{4.5}$$

$d(., \varepsilon)$ is called discrimination intensity function and ε is called discrimination factor. $d(r, \varepsilon)$ determines the minimum difference between the weight of the rth priority and the $(r + 1)$th priority. Thus, the linear model presented by Cook and Kress [1] is model (4.6).

$$Z_p = \text{Max} \sum_{r=1}^{k} w_r v_{rp}$$

$$s.t. \sum_{r=1}^{k} w_r v_{rj} \leq 1, \quad j = 1, \ldots, n \quad \text{(a)}$$
$$w_r - w_{r+1} \geq d(r, \varepsilon), \quad r = 1, \ldots, k - 1 \quad \text{(b)} \tag{4.6}$$
$$w_k \geq d(k, \varepsilon) \quad \text{(c)}$$

Model (4.6) is similar to models (2.35) and (2.44). First, it is like performance evaluation models in DEA without explicit input, and secondly, it is like models with special weight restrictions. In fact, the model (4.6) is the same as the model (2.30), in which, in addition to considering the number of votes in each position as an output, a fixed input with a value of one is also considered for all candidates. This is explained in Sect. 4.2. In other words, the result will be the same in both cases, i.e., in the case where the input value is considered to be one and in the case where the input is constant, the result will be the same. In model (4.6), each candidate chooses the best weight vector for his voting priorities, and in the optimality, one of the constraints of category (a) always holds. Constraints (b) and (c) also show the order between the weights of voting priorities. Any candidate that has an objective function value equal to one in the model (4.6) is efficient. There are several important points in this model. First, because a separate model is solved for each candidate and this model chooses the best weight vector for each candidate, it is sometimes not able to rank the candidates, because it may be $Z_p = 1$ for several candidates at the same time and create a tie in the ranking provided. Second, changing the weight difference between voting priorities in that model can change the winning candidate.

For further discussion on this model, its dual is considered. In order to present the dual model (4.6) more easily, first, the equivalent form (4.7) is considered for which the corresponding dual variables of each constraint are displayed next to them.

$$Z_p = \text{Max} \sum_{r=1}^{k} w_r v_{rp}$$

$$s.t. \sum_{r=1}^{k} w_r v_{rj} \leq 1, \ j = 1, \ldots, n \qquad \rightarrow \lambda_j$$
$$w_{r+1} - w_r \qquad \leq -d(r, \varepsilon), r = 1, \ldots, k - 1 \rightarrow s_r \tag{4.7}$$
$$-w_k \qquad \leq -d(k, \varepsilon) \qquad \rightarrow s_k$$

Thus, the dual of model (4.7) is in the form of model (4.8).

$$Z_p = \text{Min} \sum_{j=1}^{n} \lambda_j - \sum_{r=1}^{k} s_r d(r, \varepsilon)$$

$$s.t. \sum_{r=1}^{k} \lambda_j v_{1j} - s_1 = v_{1p}$$

$$\sum_{r=1}^{k} \lambda_j v_{rj} + s_{r-1} - s_r = v_{rp}, \quad r = 2, \ldots, k \quad\quad (4.8)$$

$$\lambda_j \geq 0, \qquad\qquad j = 1, \ldots, n$$

$$s_r \geq 0, \qquad\qquad r = 1, \ldots, k$$

By calculating s_1, s_2 and placing the calculated value in the objective function and constraints, we have:

$$s_1 = \sum_{r=1}^{k} \lambda_j v_{1j} - v_{1p},$$

$$\sum_{r=1}^{k} \lambda_j v_{2j} + s_1 - s_2 = v_{2p} \Rightarrow \sum_{r=1}^{k} \lambda_j v_{2j} + \left(\sum_{r=1}^{k} \lambda_j v_{1j} - v_{1p} \right) s_2 = v_{2p} \Rightarrow$$

$$\sum_{r=1}^{k} \lambda_j (v_{2j} + v_{1j}) - s_2 = v_{2p} + v_{1p} \Rightarrow s_2 = \sum_{r=1}^{k} \lambda_j (v_{2j} + v_{1j}) - (v_{2p} + v_{1p})$$

$$(4.9)$$

In the same way, by continuing the process presented in (4.9), other variables s_r can be calculated as (4.10).

$$\begin{cases} s_r = \left(\sum_{r=1}^{k} \lambda_j \left(\sum_{t=1}^{r} v_{tj} \right) \right) - \left(\sum_{t=1}^{r} v_{tp} \right) \\ \left(\sum_{r=1}^{k} \lambda_j \left(\sum_{t=1}^{r} v_{tj} \right) \right) - \left(\sum_{t=1}^{r} v_{tp} \right) \geq 0 \end{cases} \quad (4.10)$$

Considering relations (4.9) and (4.10) in model (4.8), model (4.11) is obtained.

$$Z_p = \text{Min} \sum_{j=1}^{n} \lambda_j - \sum_{r=1}^{k} \left(\sum_{j=1}^{n} \lambda_j \left(\sum_{t=1}^{r} v_{tj} \right) - \left(\sum_{t=1}^{r} v_{tp} \right) \right) d(r, \varepsilon)$$

$$s.t. \sum_{j=1}^{n} \lambda_j \left(\sum_{t=1}^{r} v_{tj} \right) \geq \left(\sum_{t=1}^{r} v_{tp} \right), \quad r = 1, \ldots, k \quad\quad (4.11)$$

$$\lambda_j \geq 0, \qquad\qquad j = 1, \ldots, n$$

Model (4.11) can be used as an envelopment model for voting.

Therefore, just as the advantages of that attitude can be used by using an attitude, the disadvantages will also show themselves. The first point mentioned above is that the simultaneous existence of several efficient candidates is one of the problems of the optimistic attitude in determining the weights. This issue has caused many methods to be invented in the DEA for ranking the DMU. The same issue also exists in the preferential voting, which is discussed in the Chap. 5. But the second point that was mentioned above, i.e., changes in the discrimination intensity function, can be called one of the most important and popular topics in preferential voting.

4.3 The Discrimination Intensity Functions

An initial proposal for the difference between the weights of priorities is called the discrimination intensity function. This function, which is determined in interaction with decision-makers, specifies the minimum difference between the weights of voting priorities. The general structure of this function is as (4.12).

$$d(r, \varepsilon) : \mathbb{N} \times \mathbb{R}^+ \to \mathbb{R}^+ \tag{4.12}$$

$d(r, \varepsilon)$ is a non-negative and non-decreasing function in terms of ε. Based on the classifications made for weight restrictions in DEA [4], the category of restrictions (4.5) can be introduced as homogeneous or non-homogeneous restrictions. Homogeneous or non-homogeneous was said because it is non-negative and can be zero or a positive value. As stated in Sect. 2.5, adding a weight restriction to the model will not improve the value of the objective function, and since the model is of the maximization type, the optimal value of the objective function will either remain constant or decrease.

The first forms of the discrimination intensity function were introduced as $d(r, \varepsilon) = \varepsilon$ or $d(r, \varepsilon) = \frac{\varepsilon}{r}$ or $d(r, \varepsilon) = \frac{\varepsilon}{r!}$. Naturally, the form of the discrimination intensity function has a direct effect on the overall structure of the weight restriction and on the feasible region of the model. Green et al. [5] considered the number of cumulative standing for each candidate, $V_{rq} = \sum_{k=1}^{r} v_{kq}$ where v_{kq} is the number of votes for candidate q in the kth priority. For example, for candidate q with four priorities, the total votes will be as (4.13):

$$V_{1q} = v_{1q}$$
$$V_{2q} = v_{1q} + v_{2q}$$
$$V_{3q} = v_{1q} + v_{2q} + v_{3q}$$
$$V_{4q} = v_{q1} + v_{2q} + v_{3q} + v_{4q} \tag{4.13}$$

Using cumulation on the number of votes sorts the weights weakly. Therefore, there will be no need to write the restrictions of the assurance region in the model, and the model (4.6) becomes as (4.14).

$$Z_p = \text{Max} \sum_{r=1}^{k} W_r V_{rp}$$

$$s.t. \quad \sum_{r=1}^{k} W_r V_{rj} \leq 1, \ j = 1, \ldots, n \qquad (4.14)$$

$$W_r \geq 0, \qquad r = 1, \ldots, k$$

where $w_r = \sum_{k=r}^{r} W_k$. For example, the objective function for candidate q with four priorities is as (4.14).

$$
\begin{aligned}
Z_p &= W_1 V_{1q} + W_2 V_{2q} + W_3 V_{3q} + W_4 V_{4q} \\
&= W_1\left(v_{1q}\right) + W_2\left(v_{1q} + v_{2q}\right) + W_3\left(v_{1q} + v_{2q} + v_{3q}\right) \\
&\quad + W_4\left(v_{q1} + v_{2q} + v_{3q} + v_{4q}\right) \\
&= \left(W_1 + W_2 + W_3 + W_4\right)\left(v_{1q}\right) + \left(W_2 + W_3 + W_4\right)\left(v_{2q}\right) \\
&\quad + \left(W_3 + W_4\right)\left(v_{3q}\right) + \left(W_4\right)\left(v_{4q}\right)
\end{aligned}
\qquad (4.15)
$$

In other words, weak order (4.16) is applied in this model.

$$
\begin{aligned}
\left(W_1 + W_2 + W_3 + W_4\right) &\geq \left(W_2 + W_3 + W_4\right) \\
&\geq \left(W_3 + W_4\right) \geq \left(W_4\right)
\end{aligned}
\qquad (4.16)
$$

The dual model (4.14) is in the form of (4.17).

$$Z_p = \text{Min} \sum_{j=1}^{n} \lambda_j$$

$$s.t. \quad \sum_{r=1}^{k} \lambda_j V_{rj} \geq V_{rp}, \ r = 1, \ldots, k \qquad (4.17)$$

$$\lambda_j \geq 0, \qquad j = 1, \ldots, n$$

As it is clear in model (4.17), this model is obtained by setting $d(r, \varepsilon) = 0$ from model (4.11). By placing $\sum_{j=1}^{n} \lambda_j = \theta$, the model (4.17) becomes (4.18).

$$Z_p = \text{Min} \ \theta$$

$$\sum_{j=1}^{n} \lambda_j = \theta$$

$$s.t. \quad \sum_{r=1}^{k} \lambda_j V_{rj} \geq V_{rp}, \ r = 1, \ldots, k \qquad (4.18)$$

$$\lambda_j \geq 0, \qquad j = 1, \ldots, n$$

The model (4.18) is similar to the structure of the model (2.3): that is, the evaluation of the efficiency of the candidate p with vector $V_{rq} = \sum_{k=1}^{r} v_{kq}$ as output and constant input vector 1. On the other hand, in model (2.3), one of the input constraints and one of the output constraints are equal in the optimal state, and the first constraint can be considered as "$\leq$". Model (4.19) is the result of considering these explanations for model (4.18).

$$Z_p = \text{Min } \theta$$

$$s.t. \quad \begin{aligned} &\sum_{j=1}^{n} \lambda_j \leq \theta \\ &\sum_{r=1}^{k} \lambda_j V_{rj} \geq V_{rp}, \quad r = 1, \ldots, k \\ &\lambda_j \geq 0, \qquad\qquad j = 1, \ldots, n \end{aligned} \tag{4.19}$$

Model (4.19) is a radial model. By considering this model and considering the structure of non-radial models in DEA, the model (4.20) can be presented as the one presented in the SBM model (2.24) to determine the efficiency of pth candidate.

$$\rho_p = \text{Min } \frac{1-s^-}{1+\frac{1}{k}\sum_{r=1}^{k}\frac{s_r^+}{R_r}}$$

$$s.t. \quad \begin{aligned} &\sum_{j=1}^{n} \lambda_j + s^- = 1 \\ &\sum_{j=1}^{n} \lambda_j y_{rj} - s_r^+ = y_{rp}, \quad r = 1, \ldots, s \\ &(S^+, S^-) \geq 0, \lambda_j \geq 0, \ j = 1, \ldots, n \end{aligned} \tag{4.20}$$

In which $R_r = \text{Max}_j\{V_{rj}\} - \text{Min}_j\{V_{rj}\}, r = 1, \ldots, k$, due to the fact that some positions are zero. Obviously, in this case, poor order is considered for voting priorities. Model (4.20) will have better performance compared to model (4.6), which will use all the inefficiency due to its non-radial nature.

A drawback of using the weak weight constraint is that two priorities may have equal weight, which is a weakness in preferential voting. We may also have a weight equal to zero, in which case the priority corresponding to that weight is not considered in voting.

Noguchi et al. [6] proposed the weight restrictions $w_1 - w_2 > w_2 - w_3 > \ldots > w_{k-1} - w_k > 0$. It means that the closer we get to the first priority, the difference between two consecutive weights should be greater, and by using it, they introduced a strong order between the weights and a fixed value for ε. Based on his idea to establish $w_{r-1} - w_r > w_r - w_{r+1}$, relations (4.21) is presented.

$$w_r - \frac{r-2}{r-1}w_{r+1} > w_r - w_{r+1} \Rightarrow w_{r-1} - w_r \geq w_r - \frac{r-2}{r-1}w_{r+1}$$

$$\underset{w_r > w_{r+1}}{\Rightarrow} \quad w_{r-1} - w_r \geq w_r - \frac{r-2}{r-1}w_r \Rightarrow (r-1)w_{r-1} \geq (r)w_r \tag{4.21}$$

In this way, the weight restrictions (4.22) were presented.

$$\begin{cases} w_1 \geq 2w_2 \geq \ldots \geq kw_k \\ w_k \geq \varepsilon = \frac{1}{(1+2+\ldots+k)L} = \frac{2}{Lk(k+1)} \end{cases} \tag{4.22}$$

where L is the total number of voters and weight restrictions (4.22) are used instead of weight restrictions in model (4.6); therefore, $\varepsilon = \frac{2}{Lk(k+1)}$ and the model is always feasible.

A few points are worth mentioning here. First, Noguchi et al.'s [6] weighted order is a strong weighted order, and secondly, due to the nature of the model (4.6), there is still the possibility of tie in the ranking. Llamazares and Pena [7] considered other conditions to establish a strong ordering between weights. They considered the relation (4.23) to construct the weight restrictions.

$$w_{r-1} - w_r \geq w_r - \alpha_{r-1}w_{r+1}$$
$$\underset{w_{r-1} > w_r}{\Rightarrow} \quad w_{r-1} \geq (2 - \alpha_{r-1})w_r \tag{4.23}$$

where $\alpha_{r-1} \in [0, 1)$. In other words, $\alpha_{r-1} = \frac{r-2}{r-1}$ and different weight orders were obtained by changing α_{r-1}. For example, if $\alpha_{r-1} = 0$, then $w_{r-1} \geq 2w_r$ and we have the relation (4.24).

$$w_1 \geq 2w_2 \geq 2^2 w_3 \geq \ldots \geq 2^{k-1} w_k \tag{4.24}$$

Based on this idea, it is possible to determine the appropriate ε, which is discussed in the next section. In general, from this idea, a strong weight order can be derived in the form of (4.25).

$$w_r - \alpha_{r-1}w_{r+1} > w_r - w_{r+1} \Rightarrow w_{r-1} - w_r \geq w_r - \alpha_{r-1}w_{r+1}$$
$$\underset{w_r > w_{r+1}}{\Rightarrow} \quad w_{r-1} - w_r \geq w_r - \alpha_{r-1}w_r \Rightarrow w_{r-1} \geq (2 - \alpha_{r-1})w_r$$
$$\Rightarrow w_1 \geq (2 - \alpha_1)w_2 \geq (2 - \alpha_1)(2 - \alpha_2)w_3 \geq \ldots \geq \prod_{i=1}^{k-1}(2 - \alpha_i)w_k \tag{4.25}$$

4.4 The Sensitivity of the Model to Epsilon

In this section, the effect of changes in the discrimination factor in model (4.6) is discussed. The general form of the discrimination intensity function, which is a bivariate function, can be separable or non-separable. Cook and Kress [1] showed that the optimal value of the objective function of model (4.6) is monotonically non-increasing with the increase of the value of ε. Therefore, the higher the value of ε, the lower the number of efficient candidates. For this purpose, the model (4.26) based on which the upper limit of the discrimination factor is determined according to the type of discrimination intensity function is presented.

$$\varepsilon^* = \text{Max} \qquad\qquad\qquad \varepsilon$$

$$\text{s.t.} \quad \sum_{r=1}^{k} w_r v_{rj} \le 1, \qquad\qquad j = 1, \ldots, n$$

$$w_r - w_{r+1} \ge d(r, \varepsilon), \ r = 1, \ldots, k-1 \qquad (4.26)$$

$$w_k \ge d(k, \varepsilon),$$

$$w_r \ge 0, \qquad\qquad\qquad r = 1, \ldots, k$$

$$\varepsilon \ge 0$$

Model (4.26) is solved considering that its constraints are the same for all candidates and the value of ε^* is obtained. At least one of the first constraints of the model (4.26) will be in "=" form in the optimality. Also, in the optimality, all the constraints related to the assurance region of the model are in "=" form. Therefore, this model is feasible for $\varepsilon \in (0, \varepsilon^*]$. It should be noted that firstly, determining the value of the overall discrimination factor of each candidate, with ε^* obtained from the model (4.26), is suitable because all candidates are evaluated with a unique weight vector. Secondly, according to the above explanations, the overall discrimination factor of each candidate is at its lowest with ε^*; and this makes it more likely to have a complete ranking. In other words, the possibility of ties in the ranking will be less.

One of the easiest ways to investigate the effects of the discrimination factor is to consider the discrimination intensity function as separable. In other words, $d(r, \varepsilon) = g(r) \times h(\varepsilon)$, where $h(\varepsilon)$ is an increasing monotonic distribution in terms of ε. In this way, relations (4.27) are presented.

$$\sum_{r=1}^{k} \sum_{l=r}^{k} d(l, \varepsilon) v_{rj} = \sum_{r=1}^{k} \sum_{l=r}^{k} g(l) h(\varepsilon) v_{rj} = h(\varepsilon) \sum_{r=1}^{k} \sum_{l=r}^{k} g(l) v_{rj} \le 1$$

$$\sum_{r=1}^{k} \sum_{l=r}^{k} g(l) v_{rj} = \pi_j$$

$$h(\varepsilon) \times \pi_j \le 1 \qquad (4.27)$$

Since the model (4.26) is to maximize ε, therefore, according to (4.28), the higher π_j is, the better the candidate's rank. So if π_p is more than all $\pi_j (j = 1, 2, \ldots, n)$, then "=" will happen on the pth constraint and in the next step the pth constraint will be removed and ε will not have a role in the ranking of the candidates.

$$\begin{cases} \text{Max } \varepsilon \\ \text{s.t.} \quad h(\varepsilon) \le \frac{1}{\pi_j}, \ j = 1, \ldots, n \end{cases} \Rightarrow \varepsilon = \min_{1 \le j \le n} \left\{ h^{-1} \left(\frac{1}{\pi_j} \right) \right\} \qquad (4.28)$$

By applying this process to the constraints of (4.22), the problem of choosing the maximum ε will be in the form of model (4.29).

$$\varepsilon^* = \text{Max} \quad \varepsilon$$

$$s.t. \quad \sum_{r=1}^{k} w_r v_{rj} \leq 1, \quad j = 1, \ldots, n$$

$$w_1 \geq 2w_2 \geq \ldots \geq kw_k$$

$$w_k \geq \varepsilon$$

$$w_r \geq 0, \quad r = 1, \ldots, k$$

$$\varepsilon \geq 0 \tag{4.29}$$

Due to the similarity of the structure of models (4.26) and (4.29), it can be easily seen that the constraints related to the assurance region are equal in the optimality. From this relations (4.30) will be established.

$$w_k = \varepsilon$$

$$kw_k = (k-1)w_{k-1} \Rightarrow w_{k-1} = \frac{k}{k-1}\varepsilon$$

$$(k-1)w_{k-1} = (k-2)w_{k-2} \Rightarrow (k-1)\frac{k}{k-1}\varepsilon$$

$$= (k-2)w_{k-2} \Rightarrow w_{k-2} = \frac{k}{k-2}\varepsilon$$

$$\vdots$$

$$w_r = \frac{k}{r}\varepsilon, \quad k = 1, \ldots, k-3 \tag{4.30}$$

According to relations (4.30), model (4.29) becomes (4.31).

$$\varepsilon^* = \text{Max} \quad \varepsilon$$

$$s.t. \quad \varepsilon \sum_{r=1}^{k} \frac{k}{r} v_{rj} \leq 1, \quad j = 1, \ldots, n$$

$$\varepsilon \geq 0 \tag{4.31}$$

Therefore, ε^* is obtained from the relation (4.32).

$$\varepsilon^* = \min_{1 \leq j \leq n} \left\{ \frac{1}{\sum_{r=1}^{k} \frac{k}{r} v_{rj}} \right\} \Leftrightarrow \varepsilon^* = \frac{1}{\max_{1 \leq j \leq n} \left\{ \sum_{r=1}^{k} \frac{k}{r} v_{rj} \right\}} \tag{4.32}$$

To provide a generalization of (4.32), it is necessary to consider the maximum value in the denominator of the fraction of this relation. For this purpose, assuming the presence of a candidate whose name was placed in the first priority by all the voters, and also assuming that the number of voters is equal to L, the general relation

(4.33) will be obtained.

$$\varepsilon^* = \frac{1}{kL} \geq \frac{2}{k(k+1)L} = \varepsilon_{Noguchi} \qquad (4.33)$$

Thus, Noguchi et al. [6] actually considered a value between 0 and ε^* for their weight order.

4.5   Examples

In this section, the preferential voting model and the structure of discrimination intensity functions are explained with a numerical example.

Suppose that 5 candidates (A, B, C, D, and F) participated in the election in a preferential voting system with 90 voters. The results of counting the ballots are shown in Table 4.1.

Table 4.2 shows the evaluation results of the candidates based on the models introduced so far.

Table 4.2 actually shows the results of implementing models (4.6) and (4.26) with two types of discrimination intensity function ($d(r, \varepsilon) = \varepsilon$ and $d(r, \varepsilon) = \frac{\varepsilon}{r}$). In these functions, the value of ε^{Max} calculated from the model (4.26) is used. The efficiency score of different candidates can be different by considering different discrimination

Table 4.1 Aggregated votes in each voting priority

Candidate	1th	2th	3th	4th	5th
A	15	15	15	10	35
B	25	10	30	15	10
C	25	15	25	25	0
D	15	25	10	25	15
E	10	25	10	15	30

Table 4.2 The results of Cook and Kress voting model

Candidate	Cook and Kress [1]			
	$d(r, \varepsilon) = \varepsilon$		$d(r, \varepsilon) = \frac{\varepsilon}{r}$	
	ε^{Max}	ODI	ε^{Max}	ODI
A	0.0033	0.9836	0.0089	1
B	0.0033	0.8033	0.0089	0.6669
C	0.0033	0.7541	0.0089	0.5633
D	0.0033	0.8852	0.0089	0.7853
E	0.0033	1	0.0089	0.9815

Table 4.3 Minimum difference between voting priorities

Priorities	$d(r, \varepsilon) = \varepsilon$	$d(r, \varepsilon) = \frac{\varepsilon}{r}$
First to second	0.0033	0.0089
Second to third	0.0033	0.00445
Third to fourth	0.0033	0.002967
Fourth to fifth	0.0033	0.002225
Fifth priority to zero	0.0033	0.00178

intensity functions. In this example, two discrimination intensity functions are used, which are used to determine the minimum difference between the weights of different voting priorities. In one, this amount is fixed and in the other, it is variable. Table 4.3 shows these values.

As can be seen in Table 4.2, candidate A is efficient with the discrimination intensity function $(d(r, \varepsilon) = \frac{\varepsilon}{r})$, but it is inefficient with the discrimination intensity function $(d(r, \varepsilon) = \varepsilon)$, and this is the opposite for candidate E. This phenomenon is due to the difference in the distance between the weights of the priorities. How to choose discrimination intensity functions depends on the type of problem. The decision maker must choose how the minimum value of the distance between the weights of the voting priorities should be set.

Table 4.4 shows the ranking of candidates with a fixed discrimination intensity function equal to zero. In the second column of this table, the model of Green et al. [5] (model 4.14) is used, and in the third column, the results of the VSBM (model 4.20) are displayed. Due to the type of discrimination intensity function and cumulative standing operation, the model (4.14) cannot perform ranking in the case that all voters have filled the ballots and the polling stations are the same as the number of candidates. In certain situations, this issue can reduce even the weight selection process in optimistic and pessimistic mode. On the other hand, in the last column of the Table 4.4, the VSBM has shown good performance with this weight restriction. Proper performance means that due to the inefficiency of model (4.14), model (4.20) shows a better result. This issue can indicate that using the structure of non-radial DEA models can be a suitable way out of preferential voting ties.

In Table 4.5, the evaluation results of the candidates presented in Table 4.1 with the model of Noguchi et al. [6] is displayed. The results of this table are presented based on $\varepsilon_{Noguchi}$ and ε^{Max}.

Table 4.4 Comparing the results of radial and non-radial models with weak order in weights

$d(r, \varepsilon) = 0$	Model (4.14)	Model (4.20)
A	1	1
B	1	0.6131
C	1	0.5753
D	1	0.7581
E	1	1

Table 4.5 The results of Noguchi voting model

Candidate	Noguchi et al. [6]			
	$\varepsilon_{Noguchi}$	ODI	ε^{Max}	ODI
A	0.00074074	1	0.0039	1
B	0.00074074	0.6818	0.0039	0.6818
C	0.00074074	0.5763	0.0039	0.5763
D	0.00074074	0.7808	0.0039	0.7808
E	0.00074074	0.9792	0.0039	0.9828

4.6   Summary

In the last three decades, special attention has been paid to preferential voting as an attitude. Perhaps more than anything, the use of DEA policy has caused this. In other words, DEA improved the performance and more use of this method in decision science in removing the shortcomings of preferential voting. The importance of the efforts of pioneers in this field can be seen in determining the assurance region of weights. The type of assurance region structure and determination of the discrimination intensity function are the most important parts of using this approach. In this chapter, different states of the assurance region and different discrimination intensity functions were discussed.

Although some researchers believe that one of the disadvantages of the models presented in this chapter is the correct diagnosis of the type of discrimination intensity function, it can be said that this is actually the flexibility of the models in achieving more reasonable results in interaction with experts. This flexibility in choosing the discrimination intensity function can definitely have different results in the ranking of candidates. If this is shown in the numerical example of this chapter.

The important point here is that many of the methods introduced in this chapter can often provide a complete ranking of candidates in real-world problems. On the other hand, the use of different methods and attitudes in DEA, such as non-radial models or voting attitudes, can also cause more differentiation between candidates, which will actually create a complete ranking. This issue was objectively demonstrated in the application of a non-radial model with cumulative position in this chapter. The authors hope that this will be an introduction to the evolution of preferential voting models based on DEA policy.

Appendix

In this section, GAMS software codes for the models described in this chapter are provided.

```
Sets
   R /R1*R5/
   J /C1*C5/;
Alias(J,L);
Alias(R,T);
Variables
      Z,W(R);
Parameters
      VP(R);
TABLE V(J,R)
          R1        R2        R3        R4        R5
C1        35        10        10        20        15
C2        10        15        30        10        25
C3         0        25        25        15        25
C4        15        25        10        25        15
C5        30        15        15        20        10;

Variables
      Z,W(R),EPSILON;
      POSITIVE VARIABLES
      W(R),EPSILON ,LAMBDA(J),S1,S2(R);
Parameters
      VP(R),EPI1,CK1(J),EPI2,CK2(J),EPINOGO,NOGOEPIMAX(J),NOGO(J)
      GREE(L),RV(R),VOSBM(J),NOGOO(J);
Equations
      ObjectiveA,CONST1A,CONST2A,CONST3A
      ObjectiveB,CONST1B,CONST2B,CONST3B
      ObjectiveC,CONST1C,CONST2C,CONST3C
      ObjectiveD,CONST1D,CONST2D,CONST3D
      ObjectiveE,CONST1E,CONST2E,CONST3E
      ObjectiveF,CONST1F,CONST2F,CONST3F
      ObjectiveG,CONST1G,CONST2G,CONST3G
      ObjectiveH,CONST1H,CONST2H,CONST3H
      ObjectiveI,CONST1I,CONST2I,CONST3I;

ObjectiveA..    Z=E=EPSILON;
CONST1A(J)..    SUM(R,W(R)*V(J,R)) =L= 1;
CONST2A(R)..    W(R)-W(R+1)       =G= EPSILON;
CONST3A..       W('R5')           =G= EPSILON;

ObjectiveB..    Z=E=SUM(R,W(R)*VP(R));
CONST1B(J)..    SUM(R,W(R)*V(J,R)) =L= 1;
CONST2B(R)..    W(R)-W(R+1)       =G= EPSILON.L;
CONST3B..       W('R5')           =G= EPSILON.L;

ObjectiveC..    Z=E=EPSILON;
CONST1C(J)..    SUM(R,W(R)*V(J,R)) =L= 1;
CONST2C(R)..    W(R)-W(R+1)       =G= EPSILON/ORD(R);
CONST3C..       W('R5')           =G= EPSILON/5;
```

```
ObjectiveD..    Z=E=SUM(R,W(R)*VP(R));
CONST1D(J)..    SUM(R,W(R)*V(J,R)) =L= 1;
CONST2D(R)..    W(R)-W(R+1)      =G= EPSILON.L/ORD(R);
CONST3D..       W('R5')          =G= EPSILON.L/5;

ObjectiveE..    Z=E=EPSILON;
CONST1E(J)..    SUM(R,W(R)*V(J,R)) =L= 1;
CONST2E(R)..    ORD(R)*W(R)      =G= (ORD(R)+1)*W(R+1);
CONST3E..       W('R5')          =G= EPSILON;

ObjectiveF..    Z=E=SUM(R,W(R)*VP(R));
CONST1F(J)..    SUM(R,W(R)*V(J,R)) =L= 1;
CONST2F(R)..    ORD(R)*W(R)      =G= (ORD(R)+1)*W(R+1);
CONST3F..       W('R5')          =G= EPSILON;

ObjectiveG..    Z=E=SUM(R,W(R)*SUM(T$(ORD(T) <= ORD(R)),VP(T)));
CONST1G(J)..    SUM(R,W(R)*SUM(T$(ORD(T) <= ORD(R)),V(J,T))) =L=
1;
CONST2G(R)..    W(R)             =G=0;

ObjectiveH..    Z=E=(1-S1)/(1+((1/5)*SUM(R,S2(R)/RV(R))));
CONST1H  ..     SUM(J,LAMBDA(J))+S1             =E=1;
CONST2H(R)..                     SUM(J,LAMBDA(J)*SUM(T$(ORD(T)   <=
ORD(R)),V(J,T)))-S2(R) =E=SUM(T$(ORD(T) <= ORD(R)),VP(T));

ObjectiveI..    Z=E=SUM(R,W(R)*VP(R));
CONST1I(J)..    SUM(R,W(R)*V(J,R)) =L= 1;
CONST2I(R)..    ORD(R)*W(R)      =G= (ORD(R)+1)*W(R+1);
CONST3I..       W('R5')          =G= (2/(5*6*90));

Model MAX_EPI1 /ObjectiveA,CONST1A,CONST2A,CONST3A /;
Model CKK1     /ObjectiveB,CONST1B,CONST2B,CONST3B /;

Model MAX_EPI2 /ObjectiveC,CONST1C,CONST2C,CONST3C /;
Model CKK2     /ObjectiveD,CONST1D,CONST2D,CONST3D /;

Model MAX_EPI_NOGO /ObjectiveE,CONST1E,CONST2E,CONST3E /;
Model NOGO_EP_MAX  /ObjectiveF,CONST1F,CONST2F,CONST3F /;

Model NOGO_ORGINAL /ObjectiveI,CONST1I,CONST2I,CONST3I /;

Model GREEN /ObjectiveG,CONST1G,CONST2G /;

Model V_SBM /ObjectiveH,CONST1H,CONST2H /;

File VOTING /Results.txt/;

Put VOTING;

LOOP(R,RV(R)=SMAX(J,V(J,R))-SMIN(J,V(J,R)));
LOOP(L,
    LOOP(R,VP(R)=V(L,R));
    Solve MAX_EPI1 Using LP Maximizing Z;
    EPI1=EPSILON.L;
    Solve CKK1      Using LP Maximizing Z;
    CK1(L)=Z.L;
    Solve MAX_EPI2     Using LP Maximizing Z;
```

```
    EPI2=EPSILON.L;
    Solve CKK2        Using LP Maximizing Z;
    CK2(L)=Z.L;
    Solve GREEN       Using LP Maximizing Z;
    GREE(L)=Z.L;
    Solve MAX_EPI_NOGO Using LP Maximizing Z;
    EPINOGO=EPSILON.L;
    Solve NOGO_EP_MAX  Using LP Maximizing Z;
    NOGOEPIMAX(L)=Z.L;
    Solve NOGO_ORGINAL Using LP Maximizing Z;
    NOGOO(L)=Z.L;
    Solve V_SBM  Using NLP MINimizing Z;
    VOSBM(L)=Z.L;
);

PUT
@10'EPI1',@19'CK1',@28'EPI2',@37'CK2',@46'GREE',
@55'E_NOGO',@64'NOGOMAX',@
73'NOGO',@82'VSBM'    /;
LOOP(L,
    PUT L.TL:7;
    PUT EPI1:8:4' ';
    PUT CK1(L):8:4' ';
    PUT EPI2:8:4' ';
    PUT CK2(L):8:4' ';
    PUT GREE(L):8:4' ';
    PUT EPINOGO:8:4' ';
    PUT NOGOEPIMAX(L):8:4' ';
    PUT NOGOO(L):8:4' ';
    PUT VOSBM(L):8:4' ';
PUT /;
);
```

References

1. Cook, W., Kress, M.: A data envelopment model for aggregating preference rankings. Manage. Sci. **36**, 1302–1310 (1990)
2. Thompson, R.G., Singleton, F.D., Thrall, R.M., Smith, B.A.: Comparative site evaluations for locating a high energy lab in Texas. Intetfaces 1380–1395 (1986)
3. Thompson, R.G., Langemeiar, L.N., Lee, C.T., Thrall, R.M.: The measurement of productive efficiency with an application to Kansas Royland wheat farming. Jesse H. Jones Graduate School of Administration, Working Paper, vol. 65 (1989)
4. Podinovski, V.V.: Production trade-offs and weight restrictions in data envelopment analysis. J. Oper. Res. Soc. **55**(12), 1311–1322 (2004)
5. Green, R., Doyle, J., Cook, W.: Preference voting and project ranking using DEA and cross-evaluation. Eur. J. Oper. Res. **90**, 461–472 (1996)
6. Noguchi, H., Ogawa, M., Ishii, H.: The appropriate total ranking method using DEA for multiple categorized purposes. J. Comput. Appl. Math. **146**, 155–166 (2002)
7. Llamazares, B., Pena, T.: Preference aggregation and DEA: an analysis of the methods proposed to discriminate efficient candidates. Eur. J. Oper. Res. **197**, 714–721 (2009)

Chapter 5
Ranking Models in Preferential Voting

Abstract Vote aggregation models in preferential voting that use DEA policy use an optimistic view in candidate evaluation. This increases the number of efficient candidates. Basic models do not allow discrimination between efficient candidates. Therefore, researchers presented ranking models from different perspectives. Some of these models focus on vote structure and some on weight restrictions. In this chapter, some of these models and ideas are presented.

5.1   Motivation to Use Ranking Models

In social systems, when a representative or representatives are to be elected through voting, and in the voting process, the votes of the candidates are tied or the quorum for election is not reached, what happens? In such cases, the election is usually extended to the second round; this means that the candidates with low votes are left out and the voting is done again among the other candidates. This process is very time consuming and expensive. In this section, models will be presented in the preferential voting process that will determine the winning candidates without re-voting.

The preferential voting model based on DEA can be a powerful and useful tool in aggregating voter votes. In this model, the interaction with the decision maker increases with the flexibility in determining the discrimination intensity function. Therefore, the end result can often satisfy the decision maker. In this model, the optimal value of the objective function is the basis for ranking candidates. But this value can be a maximum of 1. Since the classical model of preferential voting calculates this value from an optimistic point of view, it is very likely that a maximum value of 1 will be reached in various cases. Now the question arises how to distinguish between efficient candidates? In other words, how are efficient candidates ranked? Consider the following example:

Example 5.1 Cook and Kress [1] in an example considered preferential voting with six candidates and four voting priorities, which is shown in Table 5.1. 20 voters participated in this process.

© The Author(s), under exclusive license to Springer Nature Switzerland AG 2023
65
M. Soltanifar et al., *Preferential Voting and Applications: Approaches Based on Data Envelopment Analysis*, Studies in Systems, Decision and Control 471,
https://doi.org/10.1007/978-3-031-30403-3_5

Table 5.1 Aggregation of 20 voters into 6 candidates in 4 voting priorities

Candidate	1th	2th	3th	4th
a	3	3	4	3
b	4	5	5	2
c	6	2	3	2
d	6	2	2	6
e	0	4	3	4
f	1	4	3	3

Table 5.2 Efficiency scores of 6 candidates according to different epsilons

ε	a	b	c	d	e	f
0.0576923	0.7259615	0.9903846	0.9759615	1.0000000	0.4086538	0.5144231
0.0288462	0.7714844	1.0000000	1.0000000	1.0000000	0.5482272	0.6023137
0.0144231	0.7919922	1.0000000	1.0000000	1.0000000	0.6178636	0.6449069
0.0072115	0.8022461	1.0000000	1.0000000	1.0000000	0.6526818	0.6662034
0.0036058	0.8073730	1.0000000	1.0000000	1.0000000	0.6700909	0.6768517
0.0018029	0.8099365	1.0000000	1.0000000	1.0000000	0.6787954	0.6821759
0.0009014	0.8112183	1.0000000	1.0000000	1.0000000	0.6831477	0.6848379
0.0004507	0.8118591	1.0000000	1.0000000	1.0000000	0.6853239	0.6861690
0.0002254	0.8121796	1.0000000	1.0000000	1.0000000	0.6864119	0.6868345

Using the classical model of preferential voting and using the discrimination intensity function $d(r, \varepsilon) = \frac{\varepsilon}{r}$, the results of Table 5.2 are obtained. The results presented in the first row of the table are related to the maximum epsilon.

In this example, there are three efficient candidates and it is not possible to distinguish between them with the the the available information. Therefore, to eliminate this shortcoming, a process should be designed that allows distinction between candidates and provide a reasonable ranking of them. The various efforts that exist with this motive in the literature are briefly described below.

5.2 Types of Ranking Models

The classic preferential voting model allows each candidate to choose a set of weights for voting preferences, such that, that candidate be in the most favorable position relative to the other candidates. This benevolent view makes it possible to obtain multiple candidates with a maximum relative efficiency score of 1. In response, researchers have proposed ranking methods to distinguish between efficient candidates. In fact, since the classic preferential voting model is a specific DEA model for DMUs with only one input (equal to 1), many DEA ranking models have the ability to rewrite the concept of preferential voting. Each ranking method from a perspective calculates scores for ranking candidates. Some methods use super efficiency models, including the method proposed by Hashimoto [2] based on the idea of Andersen and Petersen [3]. In these models, the super efficiency of each candidate is studied with changes in the efficiency frontier after the removal of that candidate. Using the cross efficiency method [4], Green et al. [5] ranked the candidates and used an optimistic and pessimistic view of the multiple-choice set. They also suggested a weak weight order based on aggregation. Since changing the discrimination intensity function in the classic preferential voting model can lead to changing the efficiency of candidates, so some methods have been proposed based on the determination of these functions. On this basis, Noguchi et al. [6] introduced weight restrictions in a strong order and provided a way to rank candidates. Obata et al. [7] proposed a model for fair discrimination between efficient candidates, which uses weights of the same size. Foroughi et al. [8, 9] developed the Obata et al. [7] model for efficient and inefficient candidates, then reduced the computational complexity of the Obata et al. [7] model by algorithmic expression. Llamazares et al. [10] showed that by selecting different norms, results of the method of Obata et al. [7] changes. They also developed the model of Obata et al. [7] with the weight restrictions of Green et al. [5] which without solving any linear programming problem, is able to determine the winning candidate.

Wang et al. [11, 12] rated the candidates in a pessimistic manner. Soltanifar [13] stated that the efficiency range is in the [1, 0), stating that using the methods of Wang et al. [11, 12] to calculate the lower efficiency limit is not appropriate; because their method is not to determine performance in the form of DEA, then to find the lower limit of efficiency, he used a pessimistic view based on the pessimistic policy of DEA and for this purpose proposed a model with binary variables. Khodabakhshi and Aryavash [14] presented a model for ranking candidates based on optimistic and pessimistic views and according to the method of Khodabakhshi and Aryavash [15]. In general, one of the problems of this method is to get an interval for ranking, because there are different methods for ranking intervals that sometimes offer different results.

Since a set of ranking methods in DEA is based on a common set of weights, the use of this idea can also be used in ranking candidates. Hosseinzadeh et al. [16] proposed the common set of weights method for ranking DMUs in DEA, and then Liu and Peng [17] developed it. Based on this view, Wang et al. [11] proposed several models for determining the top candidate. Contreras [18] introduces a two-step method for determining the group ranking of candidates. In the first step, the weight vector model determines which candidate under evaluation has the best rank. That is, the model looks for a weight vector that minimizes the candidate rank. In the second step, a compromise answer is obtained according to the best candidate rank in the first step. Hosseinzadeh et al. [19] improved the model presented by Contreras [18], also obtained the ranking of the candidates in the worst case, and considered it as the upper limit of each candidate's group ranking. Other ranking methods in preferential voting include Llamazares [10]. Here are some of these ideas in detail.

5.3   Super Efficiency Models

Consider several high jumpers competing over an obstacle set at a certain height. Some of these athletes will not be able to reach the height of the installed obstacle and some will overcome this competition. Among the athletes who have succeeded in jumping the obstacle, which one is better? How to distinguish between the best performing athletes? One idea is to change the height of the barrier. This is the same idea used in super efficiency models. In these models, the score ceiling of each candidate (which is usually equal to 1) is removed to determine the score of the candidate who had already succeeded in reaching the predetermined ceiling (Fig. 5.1).

   Hashimoto [2] applied this idea by removing the candidate under evaluation from the constraint of the efficiency value in the voting model and presented the model (5.1). He had inspired this idea from Andersen and Petersen [3], who had used it in DEA models. In fact, in this idea, the ceiling of the efficiency value (which was equal to 1 before) is removed for the candidate under evaluation and the candidate is allowed to calculate his super-efficiency value. Certainly, theis constraint for ineffective candidates will be a redundant constraint. Therefore, the super efficiency model is only used for efficient candidates.

$$
\begin{aligned}
\text{Max} \ & \sum_{r=1}^{k} w_r v_{rp} \\
s.t. \ & \sum_{r=1}^{k} w_r v_{rj} \leq 1, \qquad j = 1, \ldots, n; \ j \neq p \\
& w_r - w_{r+1} \geq d(r, \varepsilon), \ r = 1, \ldots, k - 1 \\
& w_k \geq d(k, \varepsilon)
\end{aligned}
\tag{5.1}
$$

 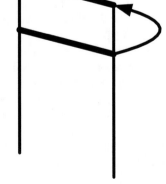

Fig. 5.1   The idea of super efficiency

By solving the model (5.1) for efficient candidates, these candidates are ranked based on their super-efficiency value. To determine the discrimination intensity functions, Hashimoto [2] also suggested to impose weight restrictions in the form of $w_r - w_{r+1} \geq w_{r+1} - w_{r+2}; r = 1, \ldots, k-2$ to the problem. With this, the difference of two consecutive weights must be greater than or equal to the next two consecutive weights. He also assumed that $d(r, \varepsilon) = \varepsilon$ (ε is a small non-Archimedean number). Model (5.2), which is known as DEA\AR exclusion model, is the result of his research.

$$\text{Max } \sum_{r=1}^{k} w_r v_{rp}$$
$$s.t. \ \sum_{r=1}^{k} w_r v_{rj} \leq 1, \qquad j = 1, \ldots, n; \ j \neq p$$
$$w_r - w_{r+1} \geq \varepsilon, \qquad r = 1, \ldots, k-1 \qquad (5.2)$$
$$w_k \geq \varepsilon$$
$$w_r - 2w_{r+1} + w_{r+2} \geq 0; \ r = 1, \ldots, k-2$$

Example 5.2 Considering the Example 5.1 and using the model (5.2), the super-efficiency value of efficient candidates is presented in the Table 5.3. In this table, the efficiency value of the ineffective candidates is also presented, and finally, the ranking of the candidates is obtained with the values provided. Although Hashimoto [2] specified the discrimination intensity functions based on the model (5.2), but in order to compare the results with what is presented in Table 5.2, the results are presented with the same discrimination intensity function $d(r, \varepsilon) = \frac{\varepsilon}{r}$. The results presented in the first row of the table are related to the maximum epsilon.

Obata and Ishii [7] proposed a model, which uses weights of the same size, to fairly discriminate between candidates. Foroughi et al. [9] extended Obata's model for efficient and inefficient candidates, then reduced the computational complexity of Obata's model with an algorithmic expression (see also [8]). Llamazares and Pena [20] showed that by choosing different norms, the winner candidate changes

Table 5.3 The results of Hashimoto's model

ε	a	b	c	d	e	f
0.0576923	0.7259615	0.9903846	0.9759615	1.0144231	0.4086538	0.5144231
0.0288462	0.7714844	1.1468531	1.0113636	1.0788899	0.5482272	0.6023137
0.0144231	0.7919922	1.2097902	1.0284091	1.1076267	0.6178636	0.6449069
0.0072115	0.8022461	1.2412587	1.0369318	1.1219952	0.6526818	0.6662034
0.0036058	0.807373	1.256993	1.0411932	1.1291794	0.6700909	0.6768517
0.0018029	0.8099365	1.2648601	1.0433239	1.1327715	0.6787954	0.6821759
0.0009014	0.8112183	1.2687937	1.0443892	1.1345676	0.6831477	0.6848379
0.0004507	0.8118591	1.2707605	1.0449219	1.1354656	0.6853239	0.686169
0.0002254	0.8121796	1.2717439	1.0451882	1.1359146	0.6864119	0.6868345

in Obata's method, they also developed Obata's model with the weight constraint of Green et al. [5], which is able to determine the winner candidate without solving any linear programming problem. Assuming $\hat{\mathbf{w}} = (\hat{w}_1, \hat{w}_2, \ldots, \hat{w}_k)$ and $\mathbf{w} = (w_1, w_2, \ldots, w_k)$, , the Obata's model is in the form of model (5.3).

$$\hat{Z}_p^* = \text{Max} \sum_{r=1}^{k} \hat{w}_r v_{rp}$$

s.t.

$$\sum_{r=1}^{k} w_r v_{rj} \leq 1, \quad j = 1, 2, \ldots, n; \ j \neq p$$

$$\sum_{r=1}^{k} w_r v_{rp} = 1$$

$$w_r - w_{r+1} \geq d(r, \varepsilon), \quad r = 1, \ldots, k-1$$

$$w_k \geq d(k, \varepsilon)$$

$$\hat{\mathbf{w}} = \alpha \mathbf{w}$$

$$\|\hat{\mathbf{w}}\| = 1$$

$$\alpha > 0 \qquad\qquad\qquad (5.3)$$

where $\|.\|$ is a certain norm. Model (5.3) is a non-linear model. According to $\|\hat{\mathbf{w}}\| = 1$ and $\hat{\mathbf{w}} = \alpha\mathbf{w}$, $\|\mathbf{w}\| = \frac{1}{\alpha}$ can be concluded. Also, since $\hat{Z}_p^* = \text{Max} \sum_{r=1}^{k} \hat{w}_r v_{rp} = \text{Max} \ \alpha \underbrace{\sum_{r=1}^{k} w_r v_{rp}}_{1}$, the objective function can be rewritten as $\frac{1}{\hat{Z}_p^*} = \text{Min} \|\mathbf{w}\|$. Thus, the model (5.3) can be rewritten as a linear model (5.4).

$$\frac{1}{\hat{Z}_p^*} = \text{Min} \ \|\mathbf{w}\|$$

s.t.

$$\sum_{r=1}^{k} w_r v_{rj} \leq 1, \qquad j = 1, 2, \ldots, n; \ j \neq p$$

$$\sum_{r=1}^{k} w_r v_{rp} = 1 \qquad\qquad\qquad (5.4)$$

$$w_r - w_{r+1} \geq d(r, \varepsilon), \ r = 1, \ldots, k-1$$

$$w_k \geq d(k, \varepsilon)$$

It is clear that the model (5.4) is not feasible for ineffective candidates and is only solved for effective candidates.

Example 5.3 Considering Example 5.1 and using model (5.4), the results of Table 5.4 are obtained for different epsilons with the discrimination intensity function $d(r, \varepsilon) =$

Table 5.4 The results of Obata's model

ε	a	b	c	d	e	f
0.0576923	Infeasible	Infeasible	Infeasible	0.2307692	Infeasible	Infeasible
0.0288462	Infeasible	0.2409856	0.2310038	0.2409856	Infeasible	Infeasible
0.0144231	Infeasible	0.2454928	0.2313555	0.2454928	Infeasible	Infeasible
0.0072115	Infeasible	0.2477464	0.2315314	0.2477464	Infeasible	Infeasible
0.0036058	Infeasible	0.2488732	0.2316194	0.2488732	Infeasible	Infeasible
0.0018029	Infeasible	0.2494366	0.2316633	0.2494366	Infeasible	Infeasible
0.0009014	Infeasible	0.2497183	0.2316853	0.2497183	Infeasible	Infeasible
0.0004507	Infeasible	0.2498591	0.2316963	0.2498591	Infeasible	Infeasible
0.0002254	Infeasible	0.2499296	0.2317018	0.2499296	Infeasible	Infeasible

$\frac{\varepsilon}{r}$. The results presented in the first row of the table are related to the maximum epsilon. As can be seen, the results for ineffective candidates are impossible.

5.4 Cross Efficiency Method

The classical model of preferential voting allows each candidate to determine the weight of voting priorities by observing the weight restrictions, in such a way that it is evaluated in the best conditions. This is the same DEA policy that the Cook and Kress [1] model is based on. The weights obtained for each candidate can also be used to evaluate other candidates. Consider a process where each candidate is evaluated not only by the best weights found for him/her, but also by the best weights found by other candidates. In other words, the candidate under evaluation will be evaluated not only by himself/herself but also by other candidates. In this way, each candidate will have an evaluation score equal to the number of all candidates. The final score of this candidate can be the average of the obtained scores. This idea, which is also presented in Fig. 5.2, is a process known as cross-efficiency evaluation. This idea was developed in the preferential voting literature by Green et al. [5]. It was inspired by what Sexton et al. [4] had used at the DEA. Based on this, first the model (4.6) is solved for each candidate and its optimal solution is obtained as $\left(w_{1p}^*, w_{2p}^*, \ldots, w_{kp}^*\right)$. Then, different evaluations are calculated for the candidate p $(p = 1, 2, \ldots, n)$ from Eq. (5.5).

$$Z_{jp} = \sum_{r=1}^{k} w_{rp}^* v_{rj}, \quad p = 1, 2, \ldots, n \qquad (5.5)$$

The final score of each candidate, which will be the basis for its ranking, is the arithmetic mean of the scores calculated for it, which is referred to in Eq. (5.6).

Fig. 5.2 The idea of cross efficiency

$$\overline{Z}_j = \frac{1}{n}\sum_{p=1}^{n} Z_{jp}, \quad j = 1, 2, \ldots n \tag{5.6}$$

Evaluating candidates with different weights has the advantage of reducing the possibility of candidate performance due to DEA's optimistic policy and makes the final score obtained for each candidate more reasonable. Therefore, the possibility of obtaining full ranking for candidates increases. Despite these advantages, the cross-efficiency method also has shortcomings that have been studied by researchers. First, solving the classical model of preferential voting may lead to different optimal solutions. These solutions may not affect the score of the candidate under evaluation, but choosing different solutions may be effective in the score of other candidates compared to the candidate under evaluation. This effect will ultimately affect the final score of the candidates and finally the final ranking. Therefore, the question arises, which one should we choose among the optimal solutions of each model? The second shortcoming of this method is why the arithmetic mean is used to aggregate the different scores for each candidate. Why arithmetic mean? Why methods such as geometric mean, winsorised are mean, InterQuartile mean-IQM, median, mode or other methods not used? In order to solve each of these shortcomings, suggestions have been provided, which will be discussed below.

5.4.1 Secondary Goal Models

The final score of each candidate in the cross-efficiency method is obtained from the average scores assigned to that candidate by all candidates. The score assigned by each candidate to the candidate under evaluation is based on the weights obtained from the model (4.6), which may have multiple optimal solutions. Choosing different solutions from among the optimal solutions of this model can be effective in the score of the candidate under evaluation. Therefore, in order to do justice between the candidates, another point of view should be presented that selects the optimal weight from among the optimal weights of a model. In the following, different views based on which a weight vector is selected from among the optimal weights of a model will be presented. These models are known as secondary goal models due to the existence of this secondary perspective.

If the model (4.6) has multiple optimal solutions, the optimal weight vector can be chosen in such a way that other candidates are evaluated in the best or worst conditions. These two views, which are called optimistic and pessimistic views respectively, were presented by Green et al. [5]. Model (5.7) is the so-called optimistic secondary goal model. In this model, among the optimal solutions of the model (4.6), a solution that is the best possible choice for other candidates is selected.

$$
\begin{aligned}
Z_{jp} = \text{Max} \sum_{r=1}^{k} w_r v_{rj}, \qquad & j = 1, 2, \ldots, n \\
s.t. \qquad \sum_{r=1}^{k} w_r v_{rp} = Z_p \\
\sum_{r=1}^{k} w_r v_{rl} \le 1, \qquad & l = 1, \ldots, n \\
w_r - w_{r+1} \ge d(r, \varepsilon), \; & r = 1, 2, \ldots, k - 1 \\
w_k \ge d(k, \varepsilon)
\end{aligned}
\tag{5.7}
$$

Model (5.8) is also known as pessimistic secondary goal model. In this model, among the optimal solutions of model (4.6), a solution that is the worst possible option for other candidates is selected. Through this model, each candidate intends to introduce a weight that is the worst possible for the evaluation of other candidates; while it has evaluated that candidate in the best conditions.

$$Z_{jp} = \text{Max} \sum_{r=1}^{k} w_r v_{rj}, \qquad j = 1, 2, \ldots, n$$

$$s.t. \quad \sum_{r=1}^{k} w_r v_{rp} = Z_p$$

$$\sum_{r=1}^{k} w_r v_{rl} \leq 1, \qquad l = 1, \ldots, n \tag{5.8}$$

$$w_r - w_{r+1} \geq d(r, \varepsilon), \ r = 1, 2, \ldots, k - 1$$

$$w_k \geq d(k, \varepsilon)$$

Optimistic and pessimistic ideas in the form presented by Green et al. [5] require solving secondary goal models with the number of candidates to determine the optimal weight. This increases the computational complexity dramatically. To reduce the computational complexity, common set of weights model can be used for two optimistic and pessimistic views. Thus, after solving the model (4.6) for each candidate, an optimistic or pessimistic common set of weights model is solved to select the optimal weight vector to evaluate other candidates. To present such a common set of weights model, considering the optimistic point of view, first the multi-objective model (5.9) is presented. In this model, among the optimal weights, the weight vector that is the best possible for other candidates is selected. By converting "max" to "min" in the objective function of this model, the pessimistic view can be presented.

$$\text{Max} \left\{ \sum_{r=1}^{k} w_r v_{rj} \right\}_{j=1, j \neq p}^{n}$$

$$s.t. \quad \sum_{r=1}^{k} w_r v_{rp} = Z_p$$

$$\sum_{r=1}^{k} w_r v_{rl} \leq 1, \qquad l = 1, \ldots, n \tag{5.9}$$

$$w_r - w_{r+1} \geq d(r, \varepsilon), \ r = 1, 2, \ldots, k - 1$$

$$w_k \geq d(k, \varepsilon)$$

This problem can be solved by converting the objective function into constraints, applying the weighting, absolute priority, goal programming, etc. This secondary goal model was presented by Soltanifar and Shahghobadi [21] and with methods such as max-sum approach, max-ordering approach, minimizing mean absolute deviation approach, led to different linear secondary goal models. They also presented other secondary goal models based on Obata and Ishii's idea [7], Liu and Peng's idea [17], and Dimitrov and Sutton's idea [22]. A new perspective that has recently been used to present secondary goal models is inspired by multi-attribute decision-making models. The secondary goal model presented by Soltanifar and Sharfi [23], which was inspired by the VIKOR method, or the secondary goal model presented by Sharfi et al. [24], inspired by the CODAS method, are such models.

5.4.2 Other Suggestions Instead of Arithmetic Mean

After calculating different scores for the candidate under evaluation, which is obtained based on the optimal weights of different candidates, the final score of the candidate should be calculated. The final score should represent other scores and be the most similar to them. In such cases, central indicators are used. The central index used in the cross-efficiency method is the arithmetic mean. But there are other central indicators such as geometric mean, winsorised are mean, InterQuartile mean-IQM, median and so on that can replace the arithmetic mean. In such cases, the decision maker can use any of them. But it should be noted that in the process of solving the problem by cross-efficiency method, a matrix called cross-efficiency matrix is formed like Table 5.5. This matrix can be considered from another point of view as a decision-making matrix in which candidates play both the role of alternatives and the role of attributes. In this way, the use of multi-attribute decision making methods can be used to evaluate candidates. The research done by Sharfi et al. [24] and Soltanifar and Sharfi [23] are of this type.

Example 5.4 Considering the Example 5.1 and using the cross-efficiency method, the matrix presented in Table 5.6 is obtained as the cross-efficiency matrix.

These results with Green's optimistic point of view and using the suitable secondary goal model are shown in Table 5.7 and with Green's pessimistic point of view and using the suitable secondary goal model in Table 5.8.

Table 5.5 Cross-efficiency matrix as a decision-matrix

		Attributes			
		Candidate 1	Candidate 2	...	Candidate n
Alternatives	Candidate 1	Z_1	Z_{12}	...	Z_{1n}
	Candidate 2	Z_{21}	Z_2	...	Z_{2n}
	⋮	⋮	⋮	...	⋮
	Candidate n	Z_{n1}	Z_{n2}	...	Z_n

Table 5.6 Green's cross efficiency matrix

	a	b	c	d	e	f
a	0.81250	0.81250	0.74390	0.50007	0.81250	0.81250
b	1.00000	1.00000	1.00000	0.66677	1.00000	1.00000
c	0.81250	0.81250	1.00000	1.00000	0.81250	0.81250
d	1.00000	1.00000	1.00000	1.00000	1.00000	1.00000
e	0.68750	0.68750	0.39024	0.00012	0.68750	0.68750
f	0.68750	0.68750	0.50000	0.16677	0.68750	0.68750

Table 5.7 Green's optimistic cross-efficiency matrix

	a	b	c	d	e	f
a	0.81250	0.81250	0.74390	0.81250	0.81250	0.81250
b	1.00000	1.00000	1.00000	1.00000	1.00000	1.00000
c	0.81250	1.00000	1.00000	1.00000	0.81250	0.81250
d	1.00000	1.00000	1.00000	1.00000	1.00000	1.00000
e	0.68750	0.68750	0.39024	0.68750	0.68750	0.68750
f	0.68750	0.68750	0.50000	0.68750	0.68750	0.68750

Table 5.8 Green's pessimistic cross-efficiency matrix

	a	b	c	d	e	f
a	0.81250	0.66667	0.50000	0.50000	0.81250	0.81250
b	1.00000	1.00000	0.66667	0.66667	1.00000	1.00000
c	0.81250	0.78571	1.00000	0.81250	0.81250	0.81250
d	1.00000	0.71429	0.95000	1.00000	1.00000	1.00000
e	0.68750	0.35000	0.00000	0.00000	0.68750	0.68750
f	0.68750	0.47500	0.16667	0.16667	0.68750	0.68750

Table 5.9 The results of cross-efficiency method

	Green	Green's pessimistic	Green's pessimistic
a	0.74900	0.80107	0.68403
b	0.94446	1.00000	0.88889
c	0.87500	0.90625	0.83929
d	1.00000	1.00000	0.94405
e	0.52339	0.63796	0.40208
f	0.56946	0.65625	0.47847

Finally, using the arithmetic mean method, the final results are presented in Table 5.9. Of course, these results could be obtained by other methods mentioned in this section.

5.5   The Idea of Contreras

Contreras [18] presented a two-step method to determine a group-ranking for candidates. In the first step of the presented method, weight vector is determined under which the candidate under evaluation has the best rank. In the second stage, an accommodating solution is attained based on the optimum ranking of the candidate in the first stage. The method is presented as follows.

Step 1 (rank optimization): To achieve the minimum rank (rank optimization) for the candidate under evaluation p ($p = 1, 2, ..., n$), model (5.10) is solved.

$$\text{Min } r_p^p$$

$$s.t. \sum_{j=1}^{k} w_j v_{ij} - \sum_{j=1}^{k} w_j v_{hj} + \delta_{ih}^p M \geq 0, \quad i \neq h$$

$$\delta_{ih}^p + \delta_{hi}^p = 1, \qquad\qquad i \neq h$$

$$\delta_{ih}^p + \delta_{hk}^p + \delta_{ki}^p > 1, \qquad i \neq h \neq k \qquad\qquad (5.10)$$

$$r_i^p = 1 + \sum_{i \neq h} \delta_{ih}^p, \qquad\qquad i = 1, \dots, n$$

$$w_j - w_{j+1} \geq d(j, \varepsilon), \qquad j = 1, \dots, k-1$$

$$w_k \geq d(k, \varepsilon) \qquad\qquad \delta_{ih}^p \in \{0, 1\}(i, j = 1, ...n)$$

where r_i^p represents the rank of candidate i in the evaluation of candidate p and M is a large enough number. In this model, in the first constraint category, if $\delta_{ih}^p = 0$, then $\sum_{j=1}^{k} w_j^p v_{ij} - \sum_{j=1}^{k} w_j^p v_{hj} \geq 0$, and therefore $\sum_{j=1}^{k} w_j^p v_{ij} \geq \sum_{j=1}^{k} w_j^p v_{hj}$. That is, candidate i is better than candidate h. In other words, rank of candidate i is lower than rank of candidate h. This model actually evaluates candidates two by two. For $i = 1$, by changing the index h ($h = 2, ..., n$), it compares candidate i with all candidate hs and determines which candidate has a better rank. This process is repeated for all candidates ($i = 1, ..., n$). So, in general, for a candidate under evaluation p, if $\delta_{ph}^P = 0$, then candidate h is better (lower rank) than candidate p. Therefore, when the rank of candidate p is to be obtained, it is enough to count the number of candidates who have a rank better than candidate p. Therefore, the rank of candidate p is equal to $r_p^p = 1 + \sum_{\substack{h=1 \\ h \neq p}}^{n} \delta_{ph}^P$, and finally the purpose is to minimize the rank of the evaluated candidate.

The second and third sets of constraints ensure that no two candidates have the same rank. In fact, constraint $\delta_{ih}^p + \delta_{hi}^p = 1$ says that it is not possible for candidate i to be superior to candidate h and candidate h to be better than candidate i. Table 5.10 shows all the cases of the third constraints category ($\delta_{ih}^o + \delta_{hk}^o + \delta_{ki}^o > 1$).

In fact, the fourth constraint category determines the rank of this candidate by counting all the candidates who have a better rank than the ith candidate.

By solving the above proposed model for each candidate, in addition to determining a rank for the candidate under evaluation, according to the fourth constraint category, a ranking vector is obtained for all candidates. Therefore, by evaluating the candidate p, the rank $R^p = (r_1^p, \dots, r_p^p, \dots, r_n^p)$ is obtained. That is, by solving the model (5.10) each time, a ranking vector is obtained, that is, there are n candidates with n different ranking vectors. Now the question is, which rank is acceptable? In this method, an ideal vector is introduced. In other words, for each candidate, the rank that has the smallest distance to his ideal rank is accepted.

Step 2 (a common set of weights): The ideal rank vector is equal to $R^I = (r_1^I, \dots, r_p^I, \dots, r_n^I)$, where $r_p^I = r_p^p$. The multi-objective linear integer model (5.11) is presented to find the rank of the candidates.

Table 5.10 Possible states for three assumed candidates

$\delta^o_{ki}=1$	$\delta^o_{hk}=1$	$\delta^o_{ih}=1$	$\delta^o_{ki}=1$	$\delta^o_{hk}=1$	$\delta^o_{ih}=0$
$x_i<x_k$	$x_k<x_h$	$x_h<x_i$	$x_i<x_k$	$x_k<x_h$	$x_i<x_h$
It is not possible			$x_i<x_k<x_h$		
$\delta^o_{ki}=0$	$\delta^o_{hk}=1$	$\delta^o_{ih}=1$	$\delta^o_{ki}=0$	$\delta^o_{hk}=1$	$\delta^o_{ih}=0$
$x_k<x_i$	$x_k<x_h$	$x_h<x_i$	$x_k<x_i$	$x_k<x_h$	$x_i<x_h$
$x_k<x_h<x_i$			$x_k<x_i<x_h$ (This state is not considered in this model)		
$\delta^o_{ki}=1$	$\delta^o_{hk}=0$	$\delta^o_{ih}=1$	$\delta^o_{ki}=1$	$\delta^o_{hk}=0$	$\delta^o_{ih}=0$
$x_i<x_k$	$x_h<x_k$	$x_h<x_i$	$x_i<x_k$	$x_h<x_k$	$x_i<x_h$
$x_h<x_i<x_k$			$x_i<x_h<x_k$ (This state is not considered in this model)		
$\delta^o_{ki}=0$	$\delta^o_{hk}=0$	$\delta^o_{ih}=1$	$\delta^o_{ki}=0$	$\delta^o_{hk}=0$	$\delta^o_{ih}=0$
$x_k<x_i$	$x_h<x_k$	$x_h<x_i$	$x_k<x_i$	$x_h<x_k$	$x_i<x_h$
$x_h<x_k<x_i$ (This state is not considered in this model)			It is not possible		

$$\text{Min } \{\alpha_i\}_{i=1}^n$$
$$s.t. \quad \sum_{j=1}^k w_j v_{ij} - \sum_{j=1}^k w_j v_{hj} + \delta_{ih} M \geq 0, \quad i \neq h$$
$$\delta_{ih} + \delta_{hi} = 1, \qquad\qquad\qquad i \neq h$$
$$\delta_{ih} + \delta_{hk} + \delta_{ki} > 1, \qquad\qquad i \neq h \neq k$$
$$r_i^G = 1 + \sum_{i \neq h} \delta_{ih}, \qquad\qquad i = 1, \ldots, n \qquad (5.11)$$
$$r_i^G - \alpha_i = r_i^i, \qquad\qquad\qquad i = 1, \ldots, n$$
$$w_j - w_{j+1} \geq d(j, \varepsilon), \qquad\qquad j = 1, \ldots, k-1$$
$$w_k \geq d(k, \varepsilon)$$
$$\delta_{ih} \in \{0, 1\} \qquad\qquad\qquad i, j = 1, \ldots n$$

Model (5.11) minimizes the distance to the ideal rank for all candidates simultaneously. The weight (relative importance) of all candidates is considered equal to one. Because the model is a multi-objective problem, different methods can be used to convert it into a single-objective problem. For example, model (5.12) is obtained from the weighting method and model (5.13) from min–max method.

$$\text{Min} \sum_{i=1}^{n} \alpha_i$$

$$s.t. \quad \sum_{j=1}^{k} w_j v_{ij} - \sum_{j=1}^{k} w_j v_{hj} + \delta_{ih} M \geq 0, \, i \neq h$$

$$\begin{array}{ll}
\delta_{ih} + \delta_{hi} = 1, & i \neq h \\
\delta_{ih} + \delta_{hk} + \delta_{ki} > 1, & i \neq h \neq k \\
r_i^G = 1 + \sum_{i \neq h} \delta_{ih}, & i = 1, \ldots, n \\
r_i^G - \alpha_i = r_i^i, & i = 1, \ldots, n \\
w_j - w_{j+1} \geq d(j, \varepsilon), & j = 1, \ldots, k-1 \\
w_k \geq d(k, \varepsilon) \\
\delta_{ih} \in \{0, 1\}
\end{array}$$

$$(5.12)$$

$$\text{Min} \max_{1 \leq i \leq n} \alpha_i$$

$$s.t. \quad \sum_{j=1}^{k} w_j v_{ij} - \sum_{j=1}^{k} w_j v_{hj} + \delta_{ih} M \geq 0, \, i \neq h$$

$$\begin{array}{ll}
\delta_{ih} + \delta_{hi} = 1, & i \neq h \\
\delta_{ih} + \delta_{hk} + \delta_{ki} > 1, & i \neq h \neq k \\
r_i^G = 1 + \sum_{i \neq h} \delta_{ih}, & i = 1, \ldots, n \\
r_i^G - \alpha_i = r_i^i, & i = 1, \ldots, n \\
w_j - w_{j+1} \geq d(j, \varepsilon), & j = 1, \ldots, k-1 \\
w_k \geq d(k, \varepsilon) \\
\delta_{ih} \in \{0, 1\} & i, j = 1, \ldots n
\end{array}$$

$$(5.13)$$

Additional studies regarding this idea were carried out by Hosseinzadeh Lotfi et al. [19]. First, they fixed the defect of model (5.10) in the presence of multiple optimal solutions by presenting a secondary goal model. Then they obtained the rank of the candidates in the worst case and considered it as the upper limit of the group ranking of each candidate. These measures led to a three-step method that has fixed many of the shortcomings of Contreras's method.

Example 5.5 Note that Contreras' method becomes infeasible considering Example 5.1. Therefore, models should be used that have solved the infallibility of this idea. Using Hosseinzadeh Lotfi's method [19], the results of Table 5.11 are obtained as individual ranking and the results of Table 5.12 as compromise ranking for different epsilons with the discrimination intensity function $d(r, \varepsilon) = \frac{\varepsilon}{r}$. The results presented in the first row of the table are related to the maximum epsilon.

Table 5.11 The individual ranking of Hosseinzadeh Lotfi's method [19]

ε	a	b	c	d	e	f
0.0576923	4	1	1	1	6	5
0.0288462	4	1	1	1	6	5
0.0144231	4	1	1	1	6	5
0.0072115	4	1	1	1	6	5
0.0036058	4	1	1	1	6	5
0.0018029	4	1	1	1	6	5
0.0009014	4	1	1	1	6	5
0.0004507	4	1	1	1	6	5
0.0002254	4	1	1	1	6	5

Table 5.12 The compromise ranking of Hosseinzadeh Lotfi's method [19]

ε	a	b	c	d	e	f
0.0576923	4	3	2	1	6	5
0.0288462	4	3	2	1	6	5
0.0144231	4	3	2	1	6	5
0.0072115	4	3	2	1	6	5
0.0036058	4	3	2	1	6	5
0.0018029	4	3	2	1	6	5
0.0009014	4	3	2	1	6	5
0.0004507	4	3	2	1	6	5
0.0002254	4	3	2	1	6	5

5.6   Further Ideas and Suggestions

So far, many ranking models have been presented by researchers. Some of these models are for ranking candidates in preferential voting and some for ranking DMUs in DEA, most of which can also be used for ranking candidates in preferential voting. Khodabakhshi et al. [14] presented a model for ranking candidates based on optimistic and pessimistic views and according to the method of Khodabakhshi and Aryavash [15]. In general, one of the problems of this category of methods is obtaining an interval for ranking, because there are different methods for ranking intervals that sometimes provide different results.

A large class of ranking methods in DEA is based on the common set of weights. Hosseinzadeh et al. [16] proposed the common set of weights method for ranking DMUs in DEA, and then Liu and Peng [17] developed it. Based on this attitude, Wang et al. [11] presented several models to determine the best candidate. Wang et al. [11, 12] ranked candidates in pessimistic mode. Soltanifar [13] stated that the efficiency interval is in the interval (0, 1], and stated that using the methods of Wang et al. [11, 12] are not suitable for calculating the lower limit of efficiency; because their method to determine the efficiency is not in the dominant DEA, then to find the lower limit of the efficiency interval, he used a pessimistic view based on the pessimistic policy of DEA and for this purpose he proposed a model with binary variables. In the following, Soltanifar's method is presented.

Suppose that an efficiency interval is to be provided for the candidate under evaluation p $(p = 1, 2, \ldots, n)$. Since the model (4.6) is presented based on the optimistic DEA policy, the optimal value of its objective function can be the upper limit of the efficiency interval (Z_p^U). To obtain the lower bound of the efficiency interval, pessimistic policy is used instead of optimistic DEA policy. That is, the minimum weighted total for the candidate under evaluation p, in such a way that the weight of other candidates obtained an efficiency score in range $(0, 1]$, and at least one candidate is efficient. Thus, model (5.8) was proposed to calculate the lower limit of the efficiency interval.

$$
\begin{aligned}
Z_p^L = \text{Min} \ & \sum_{r=1}^{k} w_r v_{rp} \\
s.t. \quad & \sum_{r=1}^{k} w_r v_{rj} \leq 1 \qquad\qquad j = 1, \ldots, n \\
& \sum_{r=1}^{k} w_r v_{r1} = 1 \ \ or \ \ \sum_{r=1}^{k} w_r v_{r2} = 1 \ or \ldots or \ \sum_{r=1}^{k} w_r v_{rn} = 1 \\
& w_r - w_{r+1} \geq d(r, \varepsilon), \qquad\qquad r = 1, \ldots, k-1 \\
& w_k \geq d(k, \varepsilon)
\end{aligned}
\tag{5.14}
$$

The model (5.14) can be converted into the mixed integer model (5.15) which can be solved by the methods of solving this type of problems.

$$
\begin{aligned}
Z_p^L = \text{Min} \ & \sum_{r=1}^{k} w_r v_{rp} \\
s.t. \quad & \sum_{r=1}^{k} w_r v_{rj} \leq 1, \qquad\quad j = 1, \ldots, n \\
& \sum_{r=1}^{k} w_r v_{rj} \geq 1 - M d_j, \ j = 1, \ldots, n \\
& \sum_{j=1}^{n} d_j \leq 1 - n \\
& d_j \in \{0, 1\}, \qquad\qquad j = 1, \ldots, n \\
& w_r - w_{r+1} \geq d(r, \varepsilon), \quad r = 1, \ldots, k-1 \\
& w_k \geq d(k, \varepsilon)
\end{aligned}
\tag{5.15}
$$

where M is a big positive quantity. In this way, for each candidate, an interval is provided in the form of $[Z_p^L, Z_p^U]$, $(p = 1, 2, \ldots, n)$, which can be the basis for ranking the candidates. The extent of priority of candidate p in relative to candidate q, will be computed using Eq. (5.16), which is a suitable basis for prioritizing candidates

Table 5.13 The results of Soltanifar's method

ε	a	b	c	d	e	f
0.0576923	0.7259615	0.9903846	0.9759615	1	0.4086538	0.5144231
0.0288462	0.3629808	0.4951923	0.4879808	0.5	0.2043269	0.2572115
0.0144231	0.1814904	0.2475962	0.2439904	0.25	0.1021635	0.1286058
0.0072115	0.0907452	0.1237981	0.1219952	0.125	0.0510817	0.0643029
0.0036058	0.0453726	0.061899	0.0609976	0.0625	0.0255409	0.0321514
0.0018029	0.0226863	0.0309495	0.0304988	0.03125	0.0127704	0.0160757
0.0009014	0.0113431	0.0154748	0.0152494	0.015625	0.0063852	0.0080379
0.0004507	0.0056716	0.0077374	0.0076247	0.0078125	0.0031926	0.0040189
0.0002254	0.0028358	0.0038687	0.0038123	0.0039062	0.0015963	0.0020095

[25].

$$P(p > q) = \frac{\max\left\{0, Z_p^U - Z_q^L\right\} - \max\left\{0, Z_p^L - Z_q^U\right\}}{\left(Z_p^U - Z_p^L\right) + \left(Z_q^U - Z_q^L\right)} \tag{5.16}$$

Example 5.6 Considering Example 5.1 and using Soltanifar's method, the results of Table 5.13 are obtained for different epsilons with the discrimination intensity function $d(r, \varepsilon) = \frac{\varepsilon}{r}$. The results presented in the first row of the table are related to the maximum epsilon.

5.7   Summary

In this chapter, how to deal with the ties caused by the application of preferential voting models based on DEA policy was studied. In this chapter, models were presented that can be used instead of setting up the second round of elections when the winning candidates cannot be recognized. These models have changed the attitude towards aggregated votes and try to improve the complete ranking of the candidates. It should be noted that since the presented classic voting model is a DEA model, therefore the ranking models are not limited to what was presented and most of the DEA ranking models can be rewritten for it. Also, the combined methods obtained by using multi-criteria decision-making models can be very attractive in the ranking of candidates. However, preferential voting ranking models are not only applicable in electoral systems, but can also be used as a powerful decision support tool in the group decision-making process.

Appendix

In this section, GAMS software codes for the models described in this chapter are provided.

Hashimoto's Model [2]

```
Sets
  R /R1*R4/
  J /C1*C6/
  B /0*8/
  K(J)
;

Alias(J,L);
TABLE V(J,R)
        R1      R2      R3      R4
C1      3       3       4       3
C2      4       5       5       2
C3      6       2       3       2
C4      6       2       2       6
C5      0       4       3       4
C6      1       4       3       3   ;

Variables Z,Z1,W(R),EPIV;
Parameters VP(R),N,COOK(J),EPI,PP ;
Equations
      Objective,CONST1,CONST2,CONST3
      ObjectiveA,CONST1A,CONST2A,CONST3A;

Objective..      Z=E=EPIV;
CONST1(J)..      SUM(R,W(R)*V(J,R)) =L= 1;
CONST2(R)..      W(R)-W(R+1)=G=EPIV/ORD(R);
CONST3..         W('R4')=G=EPIV/4;

ObjectiveA..     Z1=E=SUM(R,W(R)*VP(R));
CONST1A(J)$K(J)..    SUM(R,W(R)*V(J,R)) =L= 1;
CONST2A(R)..     W(R)-W(R+1)=G=EPI/ORD(R);
CONST3A..        W('R4')=G=EPI/4;

Model EPI_MAX /Objective,CONST1,CONST2,CONST3/;
Model Hashimoto /ObjectiveA,CONST1A,CONST2A,CONST3A/;
File VOTING /Results.txt/;
Put VOTING;
```

```
PUT'Hashimoto, Akihiro.'/;
PUT'A ranked voting system using a DEA/AR exclusion model: A
note.'/;
PUT'European journal of operational research 97.3 (1997): 600-
604.'/;

SOLVE EPI_MAX Using LP Maximizing Z;
PUT 'EPIMAX=' EPIV.L:12:8;
PUT/;

LOOP(B,
    EPI=EPIV.L/POWER(2,ORD(B)-1);
    PUT EPI:12:7' ';
);
PUT/;
LOOP(L,
    LOOP(J,K(J)=YES);
    K(L)=NO;
    LOOP(R,VP(R)=V(L,R));
    LOOP(B,
        EPI=EPIV.L/POWER(2,ORD(B)-1);
        SOLVE Hashimoto Using LP Maximizing Z1;
        PUT Z1.L:12:7' ';
    );
     PUT/;
);
```

Obata and Ishii's Model [7]

```
Sets
  R  /R1*R4/
  J  /C1*C6/
  B  /0*8/
  K(J),KK(J);

Alias(J,L);
Alias(J,LL);
Alias(R,T);

TABLE V(J,R)
       R1      R2      R3      R4
C1      3       3       4       3
C2      4       5       5       2
C3      6       2       3       2
C4      6       2       2       6
C5      0       4       3       4
C6      1       4       3       3    ;
```

```
Variables
     Z,Z2,Z1,EPIV,W(R);
Parameters
     VP(R),VK(R),N,COOK(J),OBATA(J),EPI,PP;

Equations
     Objective,CONST1,CONST2,CONST3
     ObjectiveA,CONST1A,CONST2A,CONST3A
     ObjectiveB,CONST1B,CONST2B,CONST3B,CONST4B

;

Objective..      Z=E=EPIV;
CONST1(J)..      SUM(R,W(R)*V(J,R)) =L= 1;
CONST2(R)..      W(R)-W(R+1)=G=EPIV/ORD(R);
CONST3..         W('R4')=G=EPIV/4;

ObjectiveA..      Z1=E=SUM(R,W(R)*VP(R));
CONST1A(J)..      SUM(R,W(R)*V(J,R)) =L= 1;
CONST2A(R)..      W(R)-W(R+1)=G=EPI/ORD(R);
CONST3A..         W('R4')=G=EPI/4;

ObjectiveB..      Z2=E=SUM(R,W(R));
CONST1B(J)$K(J).. SUM(R,W(R)*V(J,R)) =L= 1;
CONST2B..         SUM(R,W(R)*VP(R)) =E= 1;
CONST3B(R)..      W(R)-W(R+1)=G=EPI/ORD(R);
CONST4B..         W('R4')=G=EPI/4;

Model EPI_MAX /Objective,CONST1,CONST2,CONST3/;
Model COOK_MODEL    / ObjectiveA,CONST1A,CONST2A,CONST3A /;
Model OBATA_MODEL   /ObjectiveB,CONST1B,CONST2B,CONST3B,CONST4B/;

File VOTING /Results.txt/;

Put VOTING;

N=CARD(J);
SOLVE EPI_MAX Using LP Maximizing Z;
PUT 'EPIMAX=' EPIV.L:12:8;
PUT/;
```

```
LOOP(B,
   EPI=EPIV.L/POWER(2,ORD(B)-1);
   PUT EPI:12:7' ';
);
PUT/;
LOOP(L,
     LOOP(R,VP(R)=V(L,R));
     LOOP(J,K(J)=YES);
     K(L)=NO;
     LOOP(B,
          EPI=EPIV.L/POWER(2,ORD(B)-1);
          SOLVE COOK_MODEL Using LP Maximizing Z1;
          IF(Z1.L>0.9999,
          SOLVE OBATA_MODEL Using LP Maximizing Z2;
          PUT Z2.L:12:7' ';
          ELSE
          PUT   :12:7' '
          );
     );
   PUT/;
);
```

Hosseinzadeh Lotfi's Method [19]

```
Sets
  R  /R1*R4/
  J  /C1*C6/
  B  /0*8/
K1(J),K2(J);

Alias(J,L);
Alias(J,H);
Alias(J,K);

TABLE V(J,R)
         R1      R2      R3      R4
C1       3       3       4       3
C2       4       5       5       2
C3       6       2       3       2
C4       6       2       2       6
C5       0       4       3       4
C6       1       4       3       3  ;

Variables
     Z,Z1,W(R),EPIV;

POSITIVE VARIABLES
 RANK(J),ALPHA(J)
 ;
```

```
BINARY VARIABLES
    DELTA(J,L)   ;
Parameters
      VP(R),VK(R),N,COOK(J),M,RP(J,B),RPP(J),EPI,RA(J,B) ;

Equations
     Objective,CONST1,CONST2,CONST3
     ObjectiveA,CONST1A,CONST2A,CONST3A,CONST4A,CONST5A,CONST6A
     ObjectiveB,CONST1B
;

Objective..     Z=E=EPIV;
CONST1(J)..     SUM(R,W(R)*V(J,R)) =L= 1;
CONST2(R)..     W(R)-W(R+1)=G=EPIV/ORD(R);
CONST3..        W('R4')=G=EPIV/4;

ObjectiveA..              Z=E=SUM(J$K1(J),RANK(J));
CONST1A(J,H)$(ORD(J)<>ORD(H))..              SUM(R,W(R)*V(J,R))-
SUM(R,W(R)*V(H,R))+(DELTA(J,H)*M) =G= 0;
CONST2A(J,H)$(ORD(J)<>ORD(H))..   DELTA(J,H)+DELTA(H,J)=E=1;
CONST3A(J,H,K)$((ORD(J)<>ORD(H))AND    (ORD(J)<>ORD(K))    AND
(ORD(K)<>ORD(H)))..
                  DELTA(J,H)+DELTA(H,K)+DELTA(K,J) =G= 1;
CONST4A(J)..              RANK(J)=E=1+SUM(H$K2(H),DELTA(J,H));
CONST5A(R)..     W(R)-W(R+1)=G=EPI/ORD(R);
CONST6A..        W('R4')=G=EPI/4;

ObjectiveB..    Z=E=SUM(J,ALPHA(J));
CONST1B(J)..    RANK(J)-ALPHA(J)=E= RPP(J);

MODEL MAX_EPI /Objective,CONST1,CONST2,CONST3 /
Model HOSSEINZADEH_2013_A
/ObjectiveA,CONST1A,CONST2A,CONST3A,CONST4A,CONST5A,CONST6A /;
Model HOSSEINZADEH_2013_B
/ObjectiveB,CONST1B,CONST1A,CONST2A,CONST3A,CONST4A,CONST5A,
CONST6A /;

File VOTING /Results.txt/;
Put VOTING;

M=1;
N=CARD(J);

SOLVE MAX_EPI Using LP Maximizing Z;
PUT 'EPIMAX='EPIV.L:12:5' '/;
```

```
LOOP(B,
EPI=EPIV.L/POWER(2,ORD(B)-1);
PUT EPI:12:5' ';
);
PUT/;
LOOP(L,
    PUT L.TL:7;
    LOOP(J,K1(J)=NO);
    K1(L)=YES;
    LOOP(J,K2(J)=YES);
    K2(L)=NO;
        LOOP(B,
            EPI=EPIV.L/POWER(2,ORD(B)-1);
            SOLVE HOSSEINZADEH_2013_A Using MINLP Minimizing Z;
            PUT RANK.L(L):12:0' ';
            RP(L,B)=RANK.L(L);
    );

    PUT/;
);

LOOP(B,
        RPP(J)=RP(J,B)
        SOLVE HOSSEINZADEH_2013_B Using MINLP Minimizing Z;
        LOOP(J,
            RA(J,B)= RANK.L(J);
        );
);
PUT/; PUT/;
LOOP(J,
        PUT J.TL:7;
        LOOP(B, PUT RA(J,B):12:0' ');
        PUT /;
);
```

Soltanifar's Method [13]

```
Sets
  R  /R1*R4/
  J  /C1*C6/
  B/0*8/
  K(J),KK(J);

Alias(J,L);
Alias(J,LL);
Alias(R,T);
```

```
TABLE V(J,R)
        R1      R2      R3      R4
C1       3       3       4       3
C2       4       5       5       2
C3       6       2       3       2
C4       6       2       2       6
C5       0       4       3       4
C6       1       4       3       3   ;

Variables
     Z,Z1,W(R),EPIV;

BINARY VARIABLES
     D(J);
Parameters
     VP(R),VK(R),N,COOK(J),M,LOWER(J),EPI,PP;

Equations
     Objective,CONST1,CONST2,CONST3
     ObjectiveA,CONST1A,CONST2A,CONST3A
     ObjectiveB,CONST1B,CONST2B,CONST3B,CONST4B,CONST5B
;

Objective..     Z=E=EPIV;
CONST1(J)..     SUM(R,W(R)*V(J,R)) =L= 1;
CONST2(R)..     W(R)-W(R+1)=G=EPIV/ORD(R);
CONST3..        W('R4')=G=EPIV/4;

ObjectiveA..    Z1=E=SUM(R,W(R)*VP(R));
CONST1A(J)..    SUM(R,W(R)*V(J,R)) =L= 1;
CONST2A(R)..    W(R)-W(R+1)=G=EPI/ORD(R);
CONST3A..       W('R4')=G=EPI/4;

ObjectiveB..    Z1=E=SUM(R,W(R)*VP(R));
CONST1B(J)..    SUM(R,W(R)*V(J,R)) =L= 1;
CONST2B(J)..    SUM(R,W(R)*V(J,R)) =G= 1-(M*D(J));
CONST3B..       SUM(J,D(J))=L= N-1;
CONST4B(R)..    W(R)-W(R+1)=G=EPI/ORD(R);
CONST5B..       W('R4')=G=EPI/4;

Model EPI_MAX /Objective,CONST1,CONST2,CONST3/;
Model COOK_MODEL /ObjectiveA,CONST1A,CONST2A,CONST3A/;
Model SOLTANIFAR /ObjectiveB,CONST1B,CONST2B,CONST3B,CONST4B,
CONST5B /;

File VOTING /Results.txt/;
Put VOTING;
```

```
SOLVE EPI_MAX Using LP Maximizing Z;
PUT 'EPIMAX=' EPIV.L:12:8;
PUT/;
M=99999;
N=CARD(J);
EPI=0.00000001;
LOOP(B,
   EPI=EPIV.L/POWER(2,ORD(B)-1);
   PUT EPI:12:7' ';
);
PUT/;
LOOP(L,
      LOOP(R,VP(R)=V(L,R));
      LOOP(B,
           EPI=EPIV.L/POWER(2,ORD(B)-1);
           SOLVE SOLTANIFAR Using MIP MINimizing Z1;
           PUT Z1.L:12:7' ';
      );
    PUT/;
);
```

References

1. Cook, W., Kress, M.: A data envelopment model for aggregating preference rankings. Manage. Sci. **36**, 1302–1310 (1990)
2. Hashimoto, A.: A ranked voting system using a DEA/AR exclusion model: a note. Eur. J. Oper. Res. **97**, 600–604 (1997)
3. Andersen, P., Petersen, N.: A procedure for ranking efficient units in data envelopment analysis. Manage. Sci. **39**(10), 1261–1264 (1993)
4. Sexton, T.R., Silkman, R.H., Hogan, A.J.: Data envelopment analysis: critique and extensions. In: Silk, R.H. (ed.) Measuring Efficiency: An Assessment of Data Envelopment Analysis, vol. 32, pp. 73–105 (1986)
5. Green, R., Doyle, J., Cook, W.: Preference voting and project ranking using DEA and cross-evaluation. Eur. J. Oper. Res. **90**, 461–472 (1996)
6. Noguchi, H., Ogawa, M., Ishii, H.: The appropriate total ranking method using DEA for multiple categorized purposes. J. Comput. Appl. Math. **146**, 155–166 (2002)
7. Obata, T., Ishii, H.: A method for discriminating efficient candidates with ranked voting data. Eur. J. Oper. Res. **151**, 233–237 (2003)
8. Foroughi, A., Tamiz, M.: An effective total ranking model for a ranked voting system. Omega **33**, 491–496 (2005)
9. Foroughi, A.J.D., Tamiz, M.: A selection method for a preferential election. Appl. Math. Comput. **163**, 107–116 (2005)
10. Llamazares, B.: Aggregating preference rankings using an optimistic-pessimistic approach: closed-form expressions. Comput. Ind. Eng. **110**, 109–113 (2017)
11. Wang, Y., Luo, Y., Hua, Z.: Aggregating preference rankings using OWA operator weights. Inf. Sci. **177**, 3356–3363 (2007)
12. Wang, N., Yi, R., Liu, D.: A solution method to the problem proposed by Wang in voting systems. J. Comput. Appl. Math. **221**, 106–113 (2008)
13. Soltanifar, M.: Introducing an interval efficiency for each candidate in rankedvoting data using data envelopment analysis. Int. J. Soc. Syst. Sci. **3**(4), 346–361 (2011)

14. Khodabakhshi, M., Aryavash, K.: Aggregating preference rankings using an optimistic pessimistic approach. Comput. Ind. Eng. **85**, 13–16 (2015)
15. Khodabakhshi, M., Aryavash, K.: Ranking all units in data envelopment analysis. Appl. Math. Lett. **25**, 2066–2070 (2012)
16. Lotfi, F.H., Jahanshahloo, G., Memariani, A.: A method for finding common set of weights by multiple objective programming in data envelopment analysis. South West J. Pure Appl. Math. **1**, 44–54 (2000)
17. Liu, F., Peng, H.: Ranking of units on the DEA frontier with common weights. Comput. Oper. Res. **35**(5), 1624–1637 (2008)
18. Contreras, I.: A DEA-inspired procedure for the aggregation of preferences. Expert Syst. Appl. **38**, 564–570 (2011)
19. HosseinzadehLotfi, F., Rostamy-Malkhalifeh, M., Aghayi, N., GhelejBeigi, Z., Gholami, K.: An improved method for ranking alternatives in multiple criteria decision analysis. Appl. Math. Model. **37**, 25–33 (2013)
20. Llamazares, B., Pena, T.: Preference aggregation and DEA: an analysis of the methods proposed to discriminate efficient candidates. Eur. J. Oper. Res. **197**, 714–721 (2009)
21. Soltanifar, M., Shahghobadi, S.: Selecting a benevolent secondary goal model in data envelopment analysis cross-efficiency evaluation by a voting model. Socioecon. Plann. Sci. **47**(1), 65–74 (2013)
22. Dimitrov, S., Sutton, W.: Promoting symmetric weight selection in data envelopment analysis: a penalty function approach. Eur. J. Oper. Res. **200**(1), 281–288 (2010)
23. Soltanifar, M., Sharafi, H.A.: modified DEA cross efficiency method with negative data and its application in supplier selection. J. Comb. Optim. **43**, 265–296 (2022)
24. Sharafi, H., Soltanifar, M., Lotfi, F.H.: Selecting a green supplier utilizing the new fuzzy voting model and the fuzzy combinative distance-based assessment method. EURO J. Decis. Process. **10**, 100010 (2022)
25. Wang, Y.M., Yang, J.B., Xu, D.L.: A two-stage logarithmic goal programming method for generating weights from interval comparison matrices. Fuzzy Sets Syst. **152**, 475–498 (2005)

Chapter 6
Group Preferential Voting

Abstract What increases the power of a decision-making method as a decision support tool is the degree of satisfaction of its results. In many real-world issues, voters do not have an equal level of expertise. Since the classic voting models consider equal value for the votes of all the voters, it cannot give satisfactory results. This will reduce the power of preferential voting methods. In this chapter, group preferential voting methods are presented to handle the difference in the value of votes.

6.1 Motivation to Provide Group Preferential Voting Models

The preferential voting process is essentially a group decision-making process; because the decision in this process is based on the opinions of a group of experts. In classic preferential voting models, the votes of expert group members who participate as voters in the voting process have the same importance and influence in the final results. In other words, the characteristics of voters are not considered in these models. In other words, voters are different in their level of expertise, level of experience, managerial position and power of influence; but their votes have the same effect on the final results. The equality of the votes of the commons and properties is one of the shortcomings of the classical preferential voting models, for which models known as group preferential voting models are presented in the literature to solve it. Figure 6.1 shows such a group that participated in a voting process. Next, the issue of voting is presented, taking into account the unequal level of power and influence of voters.

Fig. 6.1 A group of voters with unequal levels of power

Suppose k candidates are to be selected from n candidates $\{C_1, C_2, ..., C_n\}$ and then ranked in terms of priority ($k \leq n$). Also assume that voters have unequal power and proficiency, that is, members are classified into m distinct categories, such that the votes of members of category i have higher importance than those of category $i + 1$. In addition, this system contains t voters. Table 6.1 provides a schematic representation of this voting process.

Where y_{rj}^i is the number of votes obtained by candidate j ($j = 1, 2, ..., n$) in priority r ($r = 1, 2, ..., k$) and by voters in category i ($i = 1, 2, ..., m$). In this way, if the categories of voters are considered to have the same impact of votes, by placing $v_{rj} = \sum\limits_{i=1}^{m} y_{rj}^i$, ($r = 1, 2, ..., k$; $j = 1, 2, ..., n$) and using the model (4.6), the candidates can be evaluated. In the following, some models that have evaluated

Table 6.1 Aggregation of votes in the group preferential voting system

	First category				Second category					mth category			
	C_1	C_2	...	C_n	C_1	C_2	...	C_n	...	C_1	C_2	...	C_n
1st	y_{11}^1	y_{12}^1	...	y_{1n}^1	y_{11}^2	y_{12}^2	...	y_{1n}^2	...	y_{11}^m	y_{12}^m	...	y_{1n}^m
2nd	y_{21}^1	y_{22}^1	...	y_{2n}^1	y_{21}^2	y_{22}^2	...	y_{2n}^2	...	y_{21}^m	y_{22}^m	...	y_{2n}^m
...	$\vdots$	$\vdots$	...	$\vdots$	$\vdots$	$\vdots$	...	$\vdots$	...	$\vdots$	$\vdots$	...	$\vdots$
kth	y_{k1}^1	y_{k2}^1	...	y_{kn}^1	y_{k1}^2	y_{k2}^2	...	y_{kn}^2	...	y_{k1}^m	y_{k2}^m	...	y_{kn}^m

candidates in preferential voting by considering this difference between voters are presented.

6.2 Types of Group Preferred Voting Models

In this section, various models that have been proposed to handle the preferential voting process considering different voting categories are presented.

6.2.1 Soltanifar's Idea [1]

Taking into account the problem stated in the previous section, Soltanifar [1] presented a model that has the ability to handle the unequal level of voters in the preferential voting process. In the following, his presented model is described.

Suppose $v_{rj}(r = 1, 2, ..., k; \ j = 1, 2, ..., n)$ and $q_j^i(i = 1, 2, ..., m; \ j = 1, 2, ..., n)$ are obtained by changing the variables presented in Eqs. (6.1) and (6.2). In fact, v_{rj} is the total number of votes that the jth candidate received in the rth priority by all voters in different voting categories, and q_j^i is the total number of votes that the ith category voters gave to the jth candidate in all voting priorities.

$$v_{rj} = \sum_{i=1}^{m} y_{rj}^i, \quad (r = 1, 2, ..., k; \ j = 1, 2, ..., n) \tag{6.1}$$

$$q_j^i = \sum_{r=1}^{k} y_{rj}^i, \quad (i = 1, 2, ..., m; \ j = 1, 2, ..., n) \tag{6.2}$$

It is clear that the higher value of q_j^i has a positive effect on the evaluation of the jth candidate. This value is at most equal to the number of voters (t). Therefore,

$x_{ij} = \left(t - q_j^i\right)$ is a non-negative factor, the greater of which will have a negative effect on the evaluation of the jth candidate. If we assume that the weight of the rth voting priority is w_r and the weight of the ith voting category is u_i, then the score presented in Eq. (6.3) can be a logical factor for evaluating the jth candidate.

$$Z_j = \frac{\sum_{r=1}^{k} w_r v_{rj}}{\sum_{i=1}^{m} u_i x_{ij}} \tag{6.3}$$

It is clear that the logical order $w_1 > w_2 > ... > w_k > 0$ and $u_1 > u_2 > ... > u_m > 0$ for the priorities as well as the weight of the voters must be established. Therefore, if the same optimistic policy of DEA is applied to weight determination, the model (6.4) is provided to evaluate the candidate under evaluation p ($p = 1, 2, ..., n$).

$$\text{Max } Z_p = \frac{\sum_{r=1}^{k} w_r v_{rp}}{\sum_{i=1}^{m} u_i x_{ip}}$$

$$s.t. \; Z_j = \frac{\sum_{r=1}^{k} w_r v_{rj}}{\sum_{i=1}^{m} u_i x_{ij}} \leq 1, \quad j = 1, 2, ..., n \tag{6.4}$$

$$w_r - w_{r+1} \geq d(r, \varepsilon), \quad r = 1, 2, ..., k-1$$
$$w_k \geq d(k, \varepsilon)$$
$$u_i - u_{i+1} \geq d'(i, \varepsilon'), \quad i = 1, 2, ..., m-1$$
$$u_m \geq d'(m, \varepsilon')$$

where $d(., \varepsilon)$ and $d'(., \varepsilon')$ are the discrimination intensity functions of the priorities and categories of voters, respectively. This model is a fractional model that can be converted to a linear programming model (6.5) by changing variable of Charnes and Cooper [2].

$$\text{Max } Z_p = \sum_{r=1}^{k} w_r v_{rp}$$

$$s.t. \sum_{i=1}^{m} u_i x_{ip} = 1$$

$$\sum_{r=1}^{k} w_r v_{rj} - \sum_{i=1}^{m} u_i x_{ij} \leq 0, \quad j = 1, 2, ..., n \qquad (6.5)$$

$$w_r - w_{r+1} \geq d(r, \varepsilon), \quad r = 1, 2, ..., k-1$$

$$w_k \geq d(k, \varepsilon)$$

$$u_i - u_{i+1} \geq d'(i, \varepsilon'), \quad i = 1, 2, ..., m-1$$

$$u_m \geq d'(m, \varepsilon')$$

For simplicity, Soltanifar [1] considered the discrimination intensity functions as $d(r, \varepsilon) = \varepsilon d_r$, $(r = 1, 2, ..., k)$ and $d'(i, \varepsilon') = \varepsilon' d'_i$, $(i = 1, 2, ..., m)$, in which d_r and d'_i are determined by the manager. He also rewritten some of the famous ranking models for his model and used them to rank candidates. The following is the use of the model in a case study provided by Soltanifar [1].

Example 6.1 In this example, the method presented in this section is used to rank five Iranian automobile manufacturing groups in terms of *"job independence"* and according to the opinions of 15 experts, including eight top managers and seven specialists with over 15 years of experience. Obviously, the opinions of top managers should be considered higher than other experts. The votes received by car manufacturing groups are presented in Table 6.2.

Model (6.5) is a good model for evaluating candidates in this case, which its result is presented in Table 6.3.

Table 6.2 Votes received by five automobile manufacturing groups in terms of job independence

Candidate	First place	Second place	Third place	Fourth place	Fifth place	Voter categorization
I	1	3	3	0	1	Top manager
	2	1	1	0	3	Expert
Z	2	1	1	1	3	Top manager
	2	1	2	1	1	Expert
B	1	2	1	2	2	Top manager
	2	0	2	2	1	Expert
S	2	1	2	1	0	Top manager
	1	2	2	2	0	Expert
P	2	0	0	5	1	Top manager
	0	3	0	2	2	Expert

Table 6.3 Efficiency scores obtained by the five manufacturing groups in terms of job independence

Candidate	I	Z	B	S	P
Efficiency score	1.000000	1.000000	0.999994	1.000000	0.999990

The results presented in Table 6.3 cannot provide a complete ranking of candidates. Therefore, it is necessary to use the ranking models to distinguish between efficient candidates and provide a suitable ranking for candidates.

6.2.2 Ebrahimnejad's Idea [3]

Ebrahimnejad also presented a model that handles the problem in this section. He also considered the Eqs. (6.1) and (6.2), and then, assuming the weight of the priorities and the weight of the categories as $(w_1, w_2, ..., w_k)$ and $(u_1, u_2, ..., u_m)$, presented the Eq. (6.6) as a score, to evaluate the jth candidate.

$$Z_j = \sum_{r=1}^{k} w_r v_{rj} + \sum_{i=1}^{m} u_i q_j^i \tag{6.6}$$

It is clear that the logical order $w_1 > w_2 > ... > w_k > 0$ and $u_1 > u_2 > ... > u_m > 0$ for the priorities as well as the weight of the voters must be established. Therefore, if the same optimistic policy of DEA is applied to weight determination, the model (6.7) is provided to evaluate the candidate under evaluation p $(p = 1, 2, ..., n)$.

$$\text{Max } Z_p = \sum_{r=1}^{k} w_r v_{rp} + \sum_{i=1}^{m} u_i q_p^i$$

$$s.t. \sum_{r=1}^{k} w_r v_{rj} + \sum_{i=1}^{m} u_i q_j^i \leq 1, \quad j = 1, 2, ..., n$$

$$w_r - w_{r+1} \geq d(r, \varepsilon), \quad r = 1, 2, ..., k-1 \tag{6.7}$$

$$w_k \geq d(k, \varepsilon)$$

$$u_i - u_{i+1} \geq d'(i, \varepsilon'), \quad i = 1, 2, ..., m-1$$

$$u_m \geq d'(m, \varepsilon')$$

where $d(., \varepsilon)$ and $d'(., \varepsilon')$ are the discrimination intensity functions of the priorities and categories of voters, respectively. Ebrahimnejad [3] also rewritten some of the famous ranking models for his model and used them to rank candidates. The following is the use of the model in a case study provided by Ebrahimnejad [3].

Example 6.2 In this example, a practical case is presented to explain the Ebrahim-nejad's idea. Every two years, members of the scientific committee of the Iranian Operation Research Society are selected by the the society members. Eligible members have at least 2 years of membership experience. Members are divided into continuous and associate members. The society consists of 30 continuous members and 25 associate members and they must select 4 members of the committee out of 7 candidates. The vote of continuous members is more important. According to their preferences, the votes collected are presented in Table 6.4.

Using the model (6.7) for the data presented in Table 6.3, the results of Table 6.5 are obtained.

The results presented in Table 6.5 cannot provide a complete ranking of candidates. Therefore, it is necessary to use the ranking models to distinguish between efficient candidates and provide a suitable ranking for candidates.

Table 6.4 Votes received from members of the Iranian Operation Research Society

Candidate	First place	Second place	Third place	Fourth place	Voter categorization
1	3	4	3	3	Continuous associate
	3	3	3	3	
2	2	9	2	4	Continuous associate
	6	6	6	2	
3	5	4	7	4	Continuous associate
	4	5	5	2	
4	6	5	4	6	Continuous associate
	3	4	7	3	
5	4	3	9	6	Continuous associate
	1	3	2	9	
6	5	4	4	5	Continuous associate
	7	2	0	4	
7	5	1	1	2	Continuous associate
	1	2	2	2	

Table 6.5 Efficiency scores obtained by seven candidates

Candidate	1	2	3	4	5	6	7
Efficiency score	0.67	1.00	1.00	1.00	1.00	1.00	0.52

6.2.3 Ebrahimnejad and Bagherzadeh's Idea [4]

Ebrahimnejad and Bagherzadeh [4] presented a non-linear model to evaluate the candidates in the group preferential voting process and then converted the presented model into a linear form with some changes. Considering the structure presented in Table 6.1 and placing $(w_1, w_2, ..., w_k)$ and $(u_1, u_2, ..., u_m)$ respectively as the weight of voting priorities and categories of voters, they used model (6.8) to apply optimistic DEA policy in vote aggregation to evaluate the candidate under evaluation p $(p = 1, 2, ..., n)$.

$$\text{Max } Z_p = \sum_{i=1}^{m} \sum_{r=1}^{k} w_r u_i y_{rp}^i$$

$$s.t. \sum_{i=1}^{m} \sum_{r=1}^{k} w_r u_i y_{rj}^i \leq 1, \quad j = 1, 2, ..., n$$

$$w_r - w_{r+1} \geq d(r, \varepsilon), \quad r = 1, 2, ..., k - 1$$

$$w_k \geq d(k, \varepsilon)$$

$$u_i - u_{i+1} \geq d'(i, \varepsilon'), \quad i = 1, 2, ..., m - 1$$

$$u_m \geq d'(m, \varepsilon') \tag{6.8}$$

where $d(., \varepsilon)$ and $d'(., \varepsilon')$ are the discrimination intensity functions of the priorities and categories of voters, respectively. To convert the nonlinear model (6.8) into an equivalent linear model, let $h_{ri} = w_r u_i$, $(r = 1, 2, ..., k; i = 1, 2, ..., m)$. Constraints on discrimination intensity functions must now be modified in terms of new transformations to preserve precedence between priorities and categories. For this purpose, the constraints related to the discrimination intensity functions of voting priorities and the constraints related to the discrimination intensity functions of voter categories are multiplied by u_i, $(i = 1, 2, ..., m)$ and w_r, $(r = 1, 2, ..., k)$ from the right and left, respectively. In this case, by placing $d_i(r, \varepsilon) = d(r, \varepsilon) \times u_i$ and $d'_r(i, \varepsilon') = w_r \times d'(i, \varepsilon')$, $(r = 1, 2, ...k; i = 1, 2, ..., m)$, , the linear model (6.9) is obtained.

$$\text{Max } Z_p = \sum_{i=1}^{m} \sum_{r=1}^{k} h_{ri} y_{rp}^i$$

$$s.t. \sum_{i=1}^{m} \sum_{r=1}^{k} h_{ri} y_{rj}^i \leq 1, \quad j = 1, 2, ..., n$$

$$h_{ri} - h_{(r+1)i} \geq d_i(r, \varepsilon), \quad r = 1, 2, ..., k - 1; i = 1, 2, ..., m$$

$$h_{ki} \geq d_i(k, \varepsilon), \quad i = 1, 2, ..., m$$

$$h_{ri} - h_{r(i+1)} \geq d'_r(i, \varepsilon'), \quad i = 1, 2, ..., m - 1; r = 1, 2, ..., k$$

$$h_{rm} \geq d'_r(m, \varepsilon'), \quad r = 1, 2, ..., k \tag{6.9}$$

It should be noted that $d_i(r, \varepsilon)$ and $d'_r(i, \varepsilon')$, $(r = 1, 2, ...k;\ i = 1, 2, ..., m)$ are variables in model (6.9).

Soltanifar et al. [5] implemented this idea in a simpler way and achieved a simpler model. They assumed that h_{ri}, $(r = 1, 2, ..., k;\ i = 1, 2, ..., m)$ is the weight of the rth voting priority in the ith voter category. Then they presented the model (6.10) in which $d(., \varepsilon)$ and $d'(., \varepsilon')$ are the discrimination intensity functions related to voting priorities and categories of voters.

$$\text{Max } Z_p = \sum_{i=1}^{m} \sum_{r=1}^{k} h_{ri} y_{rp}^i$$

$$s.t.\ \sum_{i=1}^{m} \sum_{r=1}^{k} h_{ri} y_{rj}^i \leq 1, \quad j = 1, 2, ..., n$$

$$h_{ri} - h_{(r+1)i} \geq d(r, \varepsilon), \quad r = 1, 2, ..., k-1;\ i = 1, 2, ..., m$$

$$h_{ki} \geq d(k, \varepsilon), \quad i = 1, 2, ..., m$$

$$h_{ri} - h_{r(i+1)} \geq d'(i, \varepsilon'), \quad i = 1, 2, ..., m-1;\ r = 1, 2, ..., k$$

$$h_{rm} \geq d'(m, \varepsilon'), \quad r = 1, 2, ..., k$$

(6.10)

It should be noted that $d(r, \varepsilon)$ and $d'(i, \varepsilon')$, $(r = 1, 2, ...k;\ i = 1, 2, ..., m)$ are not variables in model (6.10).

Example 6.3 In this example, a practical case is presented to explain the Ebrahimnejad and Bagherzadeh idea. Consider 6 candidates who are evaluated by voters who are divided into two categories. The first category of voters has more influence than the second category. The votes obtained by the candidates in this voting process are shown in Table 6.6.

Using this idea for the data presented in Table 6.6, the results of Table 6.7 are obtained.

The results presented in Table 6.7 cannot provide a complete ranking of candidates. Therefore, it is necessary to use the ranking models to distinguish between efficient candidates and provide a suitable ranking for candidates.

6.2.4 Soltanifar et al.'s Idea [6]

Soltanifar et al. [6] presented a two-stage approach to evaluate candidates in the group preferential voting process. In the first stage, they evaluated each candidate in different categories of voters, then in the second stage, they presented the final evaluation of each candidate by summing up the performance scores obtained by the candidates in different voting categories. The stages of their approach are as follows.

Stag 1: Considering the structure presented in Table 6.1, model (6.11) is presented to evaluate the assumed candidate p $(p = 1, 2, ..., n)$ in the category of voters i $(i =$

Table 6.6 Votes received from voters

Candidate	First place	Second place	Third place	Fourth place	Voter categorization
1	2	1	1	2	1st
	1	2	3	1	2nd
2	1	1	3	2	1st
	3	4	2	0	2nd
3	3	1	1	1	1st
	3	1	2	1	2nd
4	1	1	1	1	1st
	5	1	1	5	2nd
5	0	3	1	1	1st
	0	1	2	3	2nd
6	1	1	1	1	1st
	0	3	2	2	2nd

Table 6.7 Efficiency scores obtained by six candidates

Candidate	1	2	3	4	5	6
Efficiency score by $\varepsilon = 0.001$	0.8091875	1	0.8196875	1	0.6758125	0.6803125
Efficiency score by $\varepsilon = 0.00001$	0.81246687	1	0.81257188	1	0.68738313	0.68742813

$1, 2, …, m$).

$$z_p^i = \text{Max} \sum_{r=1}^{k} w_r y_{rp}^i$$

$$s.t. \sum_{r=1}^{k} w_r y_{rj}^i \leq 1, \quad j = 1, 2, …, n; \, j \neq p \qquad (6.11)$$

$$w_r - w_{r+1} \geq d(r, \varepsilon), \quad r = 1, 2, …, k - 1$$

$$w_k \geq d(k, \varepsilon)$$

where w_r is the rth priority weight and $d(., \varepsilon)$ is the discrimination intensity function of the priorities. Of course, Soltanifar et al. [6] used Noguchi et al.'s strong order [7] in their approach. In fact, this model gives the super-efficiency score of the candidate under evaluation with respect to the ith category of voters.

Stage 2: Considering the super-efficiency score obtained for the assumed candidate p ($p = 1, 2, …, n$) according to different categories of voters and also considering the weight of the ith category of voters equal to u_i, the final super-efficiency score of the assumed candidate p ($p = 1, 2, …, n$) from solving the model (6.12) is obtained.

$$Z_p = \text{Max} \sum_{i=1}^{m} u_i z_p^i$$

$$s.t. \sum_{i=1}^{m} u_i z_j^i, \quad j = 1, 2, ..., n; \ j \neq p \tag{6.12}$$

$$u_i - u_{i+1} \geq d'(i, \varepsilon'), \quad i = 1, 2, ..., m - 1$$

$$u_m \geq d'(m, \varepsilon')$$

where $d'(., \varepsilon')$ is the discrimination intensity function of the categories of voters. Of course, Soltanifar et al. [6] used Noguchi et al.'s strong order [7] in their approach. In the following, a part of the case study presented by Soltanifar et al. [6] is stated to explain the presented model.

Example 6.4 Soltanifar et al. [6] conducted a research to rank green indicators in the automotive industry. They selected the indicators of green design, green purchasing, green production, green transportation, environmental accountability, environmental management system, pollution control for selecting a green supplier using previous studies, expert opinion and using the Delphi method. 7 senior managers, 7 production managers, 9 supervisors, and 7 experts who had at least a master's degree and at least 10 years of work experience were used to collect the required data. Due to the different organizational level of the participants, different importance was taken into account for their opinions in decision making. In this way, senior managers were placed in the highest ranks and production managers, supervisors and experts were placed in the next ranks respectively. The votes obtained by voters in different categories for green indicators are presented in Table 6.8.

Using Soltanifar et al.'s idea [6] for the data presented in Table 6.8, the results of Table 6.9 are obtained.

It can be seen that the results presented in Table 6.7 are able to provide the complete ranking of the candidates.

6.3 Summary

In classical preferential voting models, the value of votes of all voters is considered the same. This is despite the fact that in many issues of the real world, voters have different levels of power, expertise and influence, and considering the same value for their opinion does not lead to logical and satisfactory results. Group preferential voting models are presented to solve such a defect, and in them, the value of the group of voters with more power and expertise will be higher. In this chapter, the proposed models for handling the group preferential voting process were presented. Of course, other concepts such as the concept of common set of weights can also be used to provide group preferential voting models [8].

Table 6.8 Aggregation of voters' votes to green indicators

Voting priorities	Senior managers							Production managers							Supervisors							Experts						
	1	2	3	4	5	6	7	1	2	3	4	5	6	7	1	2	3	4	5	6	7	1	2	3	4	5	6	7
Green design	2	3	2	0	0	0	0	0	0	0	7	0	0	0	3	0	4	2	0	0	0	1	0	0	4	2	0	0
Green purchasing	2	2	3	0	0	0	0	0	0	6	0	1	0	0	1	0	3	3	2	0	0	0	1	4	0	1	0	1
Green production	1	1	1	3	1	0	0	2	2	0	0	3	0	0	0	1	0	2	6	0	0	1	1	1	0	4	0	0
Green transportation	1	0	1	0	3	2	0	0	1	1	0	1	4	0	1	0	1	1	0	4	2	0	0	0	0	0	7	0
Environmental accountability	0	0	0	1	1	3	2	0	0	0	0	0	0	7	0	0	1	0	0	2	6	0	1	0	0	0	0	6
Environmental management system	1	0	0	2	2	1	1	5	1	0	0	0	1	0	4	2	1	0	0	2	0	5	0	0	2	0	0	0
Pollution control	0	1	0	1	0	1	4	0	3	0	0	2	2	0	0	5	0	1	1	1	1	0	4	2	1	0	0	0

Table 6.9 Efficiency scores obtained by green indicators

Candidate	Green design	Green purchasing	Green production	Green transportation	Environmental accountability	Environmental management system	Pollution control
Efficiency score	0.195479	0.148561	0.114269	0.084015	0.046304	0.345895	0.065476

Appendix

In this section, GAMS software codes for the models described in this chapter are provided.

```
Sets
    r "position"   /r1 *r2/
    j "Units"      /DMU1*DMU4/
    h  "class"     /h1*h2/
    k(j);

Alias (j,l);

Parameters
vo(h,r);


Table v(h,r,j)
          DMU1    DMU2    DMU3    DMU4
h1.r1   10      7       6       3
h1.r2   7       10      3       6
h2.r1   7       10      3       6
h2.r2   10      7       6       3
;


Variables
        z1
        z2
        z3
        w(r)  "Input weights"
        u(h)  "Output weights"
        m(h,r);
    positive variables
            w
            u
            m(h,r);

Equation
    Objective1
    Const1
    Const2
    Const3
    Const4
    Const5

    Objective2
```

```
        Const6
        Const7
        Const8
        Const9
        Const10

        Objective3
        Const11
        Const12
        Const13
        Const14
        Const15
        Const16
;

Objective1..
z1=e=sum(h,u(h)*sum(r,vo(h,r)))+sum(r,w(r)*sum(h,vo(h,r)));
Const1(j)..
sum(h,u(h)*sum(r,v(h,r,j)))+sum(r,w(r)*sum(h,v(h,r,j)))=l=1;
Const2..          u('h1')=g=2*u('h2');
const3..          u('h2')=g=0.0001;
Const4..          w('r1')=g=2*w('r2');
const5..          w('r2')=g=0.0001;

Objective2..      z2=e=sum(h,sum(r,m(h,r)*vo(h,r)));
Const6(j)..       sum(h,sum(r,m(h,r)*v(h,r,j)))=l=1;
Const7(r)..       m('h1',r)=g=2*m('h2',r);
const8(r)..       m('h2',r)=g=0.0001;
Const9(h)..       m(h,'r1')=g=2*m(h,'r2');
const10(h)..      m(h,'r2')=g=0.0001;

Objective3..      z3=e=sum(r,sum(h,w(r)*vo(h,r)));
Const11..         sum(h,u(h)*(52-sum(r,vo(h,r))))=e=1;
const12(j)..                    sum(r,sum(h,w(r)*v(h,r,j)))-sum(h,u(h)*(52-
sum(r,v(h,r,j))))=l=0;
Const13..         u('h1')=g=2*u('h2');
const14..         u('h2')=g=0.0001;
Const15..         w('r1')=g=2*w('r2');
const16..         w('r2')=g=0.0001;


Model                                    Super_CCR_I_ebrahimnezhad2012
/objective1,const1,const2,const3,const4,const5/;
Model                                    Super_CCR_I_ebrahimnezhad2016
/objective2,const6,const7,const8,const9,const10/;
Model                                    Super_CCR_I_soltanifar2016
/objective3,const11,const12,const13,const14,const15,const16/;

File    Super_CCR_I /C:\Users\user\Desktop\g. voting\gams\ebrahim neZhad
2012\Results.txt/;

Puttl Super_CCR_I 'Title ' System.title, @60 'Page ' System.page//;

Put Super_CCR_I;
```

```
Put @10'ebr-2012 ', @19'ebra-2016', @29'soltanifar2016'/;
Loop(l,
     loop(h,
     Loop(r,vo(h,r)=v(h,r,l)));
     );

     loop(j,
          k(j)=Yes;
          k(l)=No);
Solve Super_CCR_I_ebrahimnezhad2012 using LP Maximizing z1;

Put l.tl:6;
Put z1.l:12:6;
Solve Super_CCR_I_ebrahimnezhad2016 using LP Maximizing z2;
Put z2.l:12:6;
Solve Super_CCR_I_soltanifar2016 using LP Maximizing z3;
Put z3.l:12:6;
Put/;
);
```

References

1. Soltanifar, M.: A new voting model for groups with members of unequal power and proficiency. Inter. J. Indust. Math. **12**(2), 121–134 (2020)
2. Charnes, A., Cooper, W.W.: Programming with linear fractional functionals. Naval Res. Logis. Quart. **9**(3–4), 181–186 (1962)
3. Ebrahimnejad, A.: A new approach for ranking of candidates in voting systems. Opsearch **49**, 103–115 (2012)
4. Ebrahimnejad, A., Bagherzadeh, M.R.: Data envelopment analysis approach for discriminating efficient candidates in voting systems by considering the priority of voters. Hacettepe J. Math. Stat. **45**(1), 165–180 (2016)
5. Soltanifar, M., Tavana, M., Santos-Arteaga, F.J., Sharafi, H.: A hybrid multi-attribute decision-making and data envelopment analysis method for solving problems with heterogeneous attributes. Knowledge-Based Syst. (vol. submitted, 2022)
6. Soltanifar, M., Zargar, S.M., Homayounfar, M.: Green supplier selection: a hybrid group voting analytical hierarchy process approach. J. Operat. Res. Its Appl. (Appl. Math.) **19**(2), 113–132 (2022)
7. Noguchi, H., Ogawa, M., Ishii, H.: The appropriate total ranking method using DEA for multiple categorized purposes. J. Comput. Appl. Math. **146**, 155–166 (2002)
8. Sharafi, H., Lotfi, F.H., Jahanshahloo, G., Rostamy-malkhalifeh, M., Soltanifar, M., Razipour-GhalehJough, S.: Ranking of petrochemical companies using preferential voting at unequal levels of voting power through data envelopment analysis. Math. Sci. **13**(3), 287–297 (2019)

Chapter 7
Preferential Voting Based on Undesirable Voters

Abstract The idea that voters are divided into desirable and undesirable categories, and desirable category votes have a positive effect and undesirable category votes have a negative effect on candidate evaluation is a theory that may not seem very logical at first glance. But proposing this theory and presenting models for handling the preferential voting process in the presence of desirable and undesirable voters will eventually lead to the proposal of a method to solve MADM problems, which has many advantages. In this chapter, preferential voting and group preferential voting in the presence of desirable and undesirable voters are examined and models for handling such voters are presented. Then, using the presented models, a MADM method will be presented to determine the weights of alternatives and criteria, where there is no need to change the nature of the criteria in the process of normalizing the decision matrix.

7.1 Introduction and Motivation

The original preferential voting model proposed by Cook and Kress [1] is actually a DEA model where all DMUs have input equal to one. Therefore, it seems logical to theoretically generalize the issues faced by DEA models to preferential voting models. DEA models are commonly used to evaluate organizations that convert inputs into outputs in the form of homogeneous DMUs. Inputs are usually facilities that are desired by organizations and outputs are also desired by managers for production; but consider a factory where environmental pollutants are one of its outputs. Definitely, this is not the desired output of managers for production. The standard DEA models rely on the assumption that inputs are minimized and outputs are maximized. However, when some inputs or outputs are undesirable factors (e.g., pollutants or wastes), these outputs (inputs) should be reduced (increased) to improve inefficiency.

Performance evaluation in the presence of undesirable factors was first proposed by Färe et al. [2] in a non-linear DEA model. Scheel [3] proposed radial models that simultaneously considered desirable and undesirable outputs. After this, many

© The Author(s), under exclusive license to Springer Nature Switzerland AG 2023
M. Soltanifar et al., *Preferential Voting and Applications: Approaches Based on Data Envelopment Analysis*, Studies in Systems, Decision and Control 471,
https://doi.org/10.1007/978-3-031-30403-3_7

attempts were made to provide DEA models in the presence of undesirable factors [4–6].

Certainly, since in the process of preferential voting, candidates are considered as DMUs with an input equal to one, therefore, the idea of undesirable inputs is excluded for the basic model of preferential voting. But it is possible to design models in the presence of undesirable outputs. But what does such an idea mean? When the outputs are undesirable, it means that there are voters whose vote not only does not have a positive effect on the candidate's choice, but also negatively affects the candidate's choice. In other words, in the process of preferential voting, voters imagine that the candidates do not want to receive their votes because they have a negative impact on their choice. Such an idea that a group participates in the voting and their votes have a negative impact on the selection of candidates does not seem very logical in terms of the social approach of the preferential voting process. But if preferential voting models are viewed as tools for decision support, this assumption can lead to powerful tools for decision support. In this chapter, theoretically, the assumption of the existence of undesirable voters in the preferential voting process is based and the models presented in this regard are stated. Then the application of the models will be presented with numerical examples.

7.2   Preferential Voting in the Presence of Undesirable Voters

In the process of preferential voting, suppose voters are divided into two desirable and undesirable groups. Of course, they are unaware of this issue because knowing this can affect their vote. Both groups participate in the voting process and choose their desired representatives from among the candidates and present their desired preference in the form of a ballot. In the total votes of the voters, the votes of the desirable category are supposed to have a positive effect and the votes of the undesirable category are supposed to have a negative effect.

Suppose k candidates are to be selected from n candidates $\{C_1, C_2, ..., C_n\}$ and then ranked in terms of priority $(k \leq n)$. Denote by y_{rj}, $(r = 1, 2, \ldots k; j = 1, 2, \ldots, n)$, the number of votes obtained by the jth candidate in the rth position within the group of desirable voters and by x_{rj}, $(r = 1, 2, \ldots k; j = 1, 2, \ldots, n)$, the number of votes obtained by the jth candidate in the rth position within the group of undesirable voters. Table 7.1 shows how these votes are aggregated.

As the positive effect of votes of lower priorities in the category of desirable voters is greater than the positive effect of votes of higher priorities; the negative effect of lower priority votes in the category of undesirable voters will be greater than the negative effect of higher priority votes in this category. Therefor, if $(w_1, w_2, ..., w_k)$ and $(u_1, u_2, ..., u_k)$ are respectively the weight vectors of voting priorities in desirable and undesirable categories of voters, we have $w_1 > w_2 > ... > w_k$ and $u_1 > u_2 >$

Table 7.1 Aggregation of votes of desirable and undesirable voters

	Desirable				Undesirable			
Voting priorities	C_1	C_2	...	C_n	C_1	C_2	...	C_n
1st	y_{11}	y_{12}	...	y_{1n}	x_{11}	x_{12}	...	x_{1n}
2nd	y_{21}	y_{22}	...	y_{2n}	x_{21}	x_{22}	...	x_{2n}
...	$\vdots$	$\vdots$	...	$\vdots$	$\vdots$	$\vdots$	...	$\vdots$
Mth	y_{m1}	y_{m2}	...	y_{mn}	x_{m1}	x_{m2}	...	x_{mn}

$... > u_k.$ .If the DEA policy is to be used in handling this voting system, the weight vectors should be selected in such a way that the positive effect of votes of desirable voters is the greatest and the negative effect of votes of undesirable voters is the least effective in the evaluation of candidates. Model (7.1) was presented by Soltanifar and Sharfi [7] to handle the preferential voting process in the presence of undesirable voters for the assumed candidate p $(p = 1, 2, ..., n)$. Of course, another similar model has been presented by Soltanifar and Heidariyeh [8].

$$Max\, Z_p^+ = \sum_{r=1}^{k} w_r y_{rp}$$

$$Min\, Z_p^- = \sum_{r=1}^{k} u_r x_{rp}$$

$$s.t. \sum_{r=1}^{k} w_r y_{rj} \leq 1, \quad j = 1, 2, ..., n$$

$$\sum_{r=1}^{k} u_r x_{rj} \geq 1, \quad j = 1, 2, ..., n$$

$$w_r - w_{r+1} \geq d^+(r, \varepsilon), \quad r = 1, 2, ..., k - 1$$

$$w_k \geq d^+(k, \varepsilon)$$

$$u_r - u_{r+1} \geq d^-(r, \varepsilon), \quad r = 1, 2, ..., k - 1$$

$$u_k \geq d^-(k, \varepsilon)$$

(7.1)

Model (7.1) is a bi-objective linear programming model, where $d^+(., \varepsilon)$ and $d^-(., \varepsilon)$ are the discrimination intensity functions related to desirable and undesirable category voters, respectively. This model considers the strategy of DEA for desirable voters and undesirable voters simultaneously. It also makes it possible to determine the distance between voting priorities in both desirable and undesirable categories after interacting with the decision maker, and thus the final results are more reasoned and more acceptable to the decision maker. In model (7.1), if the distance between the voting priorities is the same for both desirable and undesirable voters, $d^+(., \varepsilon)$ and $d^-(., \varepsilon)$ will be the same. Also the optimal value of both efficiency scores Z_p^+ and Z_p^- is limited to 1. This model is capable of being solved by methods which resolve multi-objective problems, such as, the conversion of objective function to constraints method, weighting method, absolute priority method, and the goal programming method and like them. In utilizing the goal programming method and since, the goal of first and second objective functions are equal to 1, model (7.1) is converted to model (7.2) which is a linear programming model.

$$\text{Min } d_1 + d_2$$

$$\sum_{r=1}^{k} w_r y_{rp} + d_1 = 1$$

$$\sum_{r=1}^{k} u_r x_{rp} - d_2 = 1$$

$$\sum_{r=1}^{k} w_r y_{rj} \leq 1, \quad j = 1, 2, \ldots, n$$

$$\sum_{r=1}^{k} u_r x_{rj} \geq 1, \quad j = 1, 2, \ldots, n \qquad (7.2)$$

$$w_r - w_{r+1} \geq d^+(r, \varepsilon), \quad r = 1, 2, \ldots, k-1$$

$$w_k \geq d^+(k, \varepsilon)$$

$$u_r - u_{r+1} \geq d^-(r, \varepsilon), \quad r = 1, 2, \ldots, k-1$$

$$u_k \geq d^-(k, \varepsilon)$$

$$d_1, d_2 \geq 0$$

In model (7.2), the purpose is to minimize the deviation variables of each objective function from the goal; where the votes of the voters are used in two categories, desirable and undesirable. Now, if $\left(w_1^*, \ldots, w_k^*, u_1^*, \ldots u_k^*\right)$ is the optimal solution of the model (7.2), then Z_p^* obtained from the Eq. (7.3), can be the performance score of the hypothetical candidate p ($p = 1, 2, \ldots, n$) and be the basis for evaluating the candidates.

$$Z_p^* = \frac{Z_p^+}{Z_p^-} = \frac{\sum_{r=1}^{k} w_r^* y_{rp}}{\sum_{r=1}^{k} u_r^* x_{rp}} \qquad (7.3)$$

So far, the issue of the existence of undesirable voters has been raised theoretically and a model for handling the preferential voting process in the presence of undesirable voters has been presented. But the application of this model is still a question that needs to be answered. In the following, an application of the proposed model is presented.

As mentioned earlier, classifying voters into desirable and undesirable categories may raise the question of whether this is not a violation of voting principles. What application can this modelling have in industry and society? To clear up the ambiguities and answer the questions, we introduce this process as applied in providing a Multi-Attribute Decision-Making (MADM) method. In many MADM models, to select an alternative among a number of alternatives or to rank several alternatives by considering several criteria, we are faced with criteria of types of benefits and costs. For example, when choosing a car, quality is the criterion of profit, and when it is higher, it is more desirable. But fuel consumption is a cost measure and when it is less it will be desirable for the decision maker. In most of the models presented for these problems, cost criteria are converted into profit criteria in the normalization process, and then the ranking process takes place. The proposed voting model in the presence of undesirable voters can solve the problem without changing the nature of each criterion by considering a benefit criterion as the desirable voter and a cost criterion as the undesirable voter. In fact, the results obtained by each alternative will be effective in each benefit criterion with a positive impact and in each cost criterion with a negative impact in the ranking of that alternative. Therefore, based on the scores of each alternative in each criterion, the alternatives can be ranked in that criterion. Placement of any of the alternatives with any criterion in each voting priority is considered as a vote for that alternative in that priority. This vote can be desirable or undesirable depending on the criterion of profit or cost. Then, votes are aggregated using model (7.2).

Let us describe the proposed method. Assume we want to evaluate n similar alternatives $(A_1, A_2, \ldots, A_n)$ by considering s multiple attributes $(C_1, C_2, \ldots, C_s)$. Also suppose that some criteria are profit type and some are cost type. Equation (7.4) presents the overall decision matrix for this problem:

$$D = \begin{bmatrix} t_{11} & \cdots & t_{1s} \\ \vdots & \ddots & \vdots \\ t_{n1} & \cdots & t_{ns} \end{bmatrix} \tag{7.4}$$

Step 1: Normalize the decision matrix (7.4) using Eq. (7.5).

$$\bar{t}_{pq} = \frac{t_{pq}}{\underset{1 \leq j \leq n}{\text{Max }} t_{jq}}; \ p = 1, 2, \ldots, n; \ q = 1, 2, \ldots, s. \tag{7.5}$$

It should be noted that normalization is used to standardize the unit of measurement regardless of the attribute type.

Step 2: Assign an importance weight such as $(h_1, h_2, ..., h_s)$ to each attribute so that $\sum_{q=1}^{s} h_q = 1$.

Step 3: Calculate the weighted normalized matrix using Eq. (7.6).

$$\hat{D} = \begin{bmatrix} \hat{t}_{11} & \cdots & \hat{t}_{1s} \\ \vdots & \ddots & \vdots \\ \hat{t}_{n1} & \cdots & \hat{t}_{ns} \end{bmatrix} = \begin{bmatrix} \bar{t}_{11} & \cdots & \bar{t}_{1s} \\ \vdots & \ddots & \vdots \\ \bar{t}_{n1} & \cdots & \bar{t}_{ns} \end{bmatrix} \times \begin{bmatrix} h_1 & \cdots & 0 \\ \vdots & \ddots & \vdots \\ 0 & \cdots & h_s \end{bmatrix} \tag{7.6}$$

Step 3: Form the voting matrix shown in Table 7.1 using equations. (7.7) and (7.8). The number of candidates and voting priorities will be equal in the resulting Table 7.1 $(k = n)$ since we need to identify and allocate the priority of each alternative among all the alternatives in each category.

$$y_{rj} = \sum_{\substack{q \in \text{ category of profit attributes} \\ \hat{t}_{jq} \text{ is the } r\text{th priority in the column}}} \hat{t}_{jq} \tag{7.7}$$

$$x_{rj} = \sum_{\substack{q \in \text{ category of cost attributes} \\ \hat{t}_{jq} \text{ is the } r\text{th priority in the column}}} \hat{t}_{jq} \tag{7.8}$$

Step 4: Solve model (7.2) and calculate the score of the alternatives using Eq. (7.3). The flowchart of the proposed method is presented in Fig. 7.1.

Next, to show the applicability of the MADM method, a case study presented by Soltanifar and Sharfi [7] is presented.

Example 7.1. In this example, a case study presented by Soltanifar and Sharfi [7] is given to demonstrate the applicability of the MADM method described in this section.

A company intends to choose the best type of cutting fluid among four types A_1, A_2, A_3 and A_4 for its milling machines, according to the defined attributes. Experts defined the attributes such as the friction (C_1), stoning temperature (C_2), recycling capability (C_3), work surface roughness (C_4), and stability (C_5), and the qualitative attributes are converted into the quantitative attributes. The decision matrix is shown in Table 7.2.

The weights of the criteria are also presented as $(h_1, h_2, \cdots, h_5) = (0.331, 0.181, 0.369, 0.072, 0.047)$. Thus, the weighted normalized decision matrix can be presented in Table 7.3. Note that in the formation of Table 7.3, linear normalization (Eq. 7.5) is used.

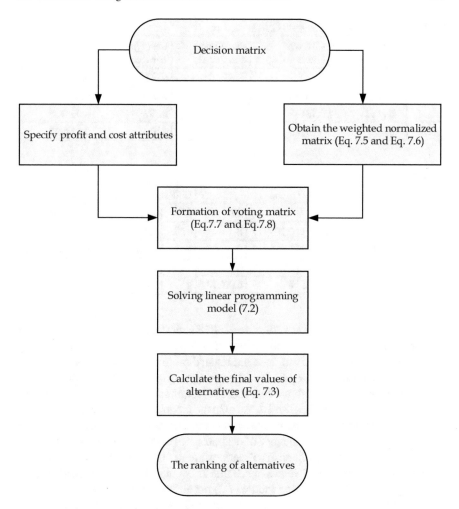

Fig. 7.1 The flowchart of the MADM method by preferential voting process

Table 7.2 The decision matrix

	Cost type criterion C_1	Cost type criterion C_2	Benefit type criterion C_3	Cost type criterion C_4	Benefit type criterion C_5
A_1	0.035	847	0.335	1.760	0.590
A_2	0.027	834	0.335	1.680	0.665
A_3	0.037	808	0.590	2.400	0.500
A_4	0.028	821	0.500	1.590	0.410

Table 7.3 The weighted normalized decision matrix

	Cost type criterion C_1	Cost type criterion C_2	Benefit type criterion C_3	Cost type criterion C_4	Benefit type criterion C_5
A_1	0.632838	0.819	0.209517	0.680533	0.041699
A_2	0.488189	0.80643	0.209517	0.6496	0.047
A_3	0.669	0.781289	0.369	0.928	0.035338
A_4	0.50627	0.79386	0.312712	0.6148	0.028977

Table 7.4 The voting matrix

Alternatives	Desirable voters (C_3 and C_5)				Undesirable voters (C_1, C_2 and C_4)			
	1st	2nd	3rd	4th	1st	2nd	3rd	4th
A_1	0	0.041699	0.209517	0	0.819	1.313371	0	0
A_2	0.047	0	0.209517	0	0	0.80643	0.6496	0.488189
A_3	0.369	0	0.035338	0	1.597	0	0	0.781289
A_4	0	0.312712	0	0.028977	0	0	1.30013	0.6148

Now we need to create the voting matrix. This matrix consists of two parts. One for desirable voters (profit criteria) and the other for undesirable voters (cost criteria). At each voting priority, the weighted aggregated scores obtained by the alternative will be placed at each voting priority. The voting matrix will be as in Table 7.4.

Applying model (7.2) to the data in Table 7.4 requires determining the discrimination intensity function $d^+(., \varepsilon)$ and $d^-(., \varepsilon)$ in the model. This function is determined after interacting with the decision maker and in fact determines the distance between the voting periorities in the category of desirable and undesirable voters. In this case study, both discrimination intensity functions are assumed equal to $\varepsilon \geq 0$ and the model is solved for different values of ε. Table 7.5 shows the score of each alternative using Eq. (7.3) and after solving model (7.2) for different values of ε.

The last column of Table 7.5 contains the results for ε_{max}; That is, ε_{max} is the maximum possible value to achieve the feasibility in model (7.2). Thus, based on the values obtained in Table 7.5, the alternatives can be ranked.

7.3 Group Preferential Voting in the Presence of Undesirable Voters

In Chap. 6, the process of preferential voting when voters are divided into categories with unequal levels of power, expertise and influence was examined and various models inspired by the ideas of different researchers were presented. This process is described in the literature as group preferential voting. In this section, the process of

Table 7.5 The score of each alternative for different values of ε

Alternatives	0.0001	0.001	0.01	0.06465478	0.12930956	0.16163696	0.21551594	0.32327391	0.64654782
A_1	0.557727364	0.555757918	0.536679969	0.440707016	0.358272797	0.325598888	0.280139553	0.208572658	0.075405
A_2	0.624760243	0.623907393	0.615456545	0.566997301	0.515339998	0.491533176	0.454500914	0.377557756	0.14430243
A_3	0.813894338	0.811979088	0.793310949	0.696119976	0.608002805	0.571811884	0.520203953	0.431472634	0.215736317
A_4	0.847423632	0.847114365	0.844021695	0.825240676	0.803023357	0.791914697	0.773400265	0.708495627	0.300807141

group preferential voting was investigated in the presence of desirable and undesirable voters. In other words, it is assumed that voters are divided into two categories, desirable and undesirable, and each category is placed in categories with different levels of power and expertise. Votes of desirable voters with a positive effect and votes of undesirable voters with a negative effect are effective in the evaluation of candidates. Also, in each desirable or undesirable category, the positive or negative impact of voters' votes placed in different priorities is different. This problem is first presented theoretically and the appropriate model to handle it is presented; then, from the presented theory and modelling, a MADM method is proposed to determine the weights of criteria and alternatives.

Let's assume that the voters are in two desirable and undesirable categories, and in each category they are divided into t sub-categories with different levels of power and expertise. The positive (negative) effect of the sth category votes from desirable (undesirable) voters is greater than the positive (negative) effect of the $(s + 1)$th category votes. Also suppose k candidates are to be selected from n candidates $\{C_1, C_2, ..., C_n\}$ and then ranked in terms of priority $(k \leq n)$. Denote by y_{rj}^s, $(s = 1, 2, ..., t \, ; \, r = 1, 2, ... k \, ; \, j = 1, 2, ..., n)$, the number of votes obtained by the jth candidate in the rth position within the group of sth desirable voters and by x_{rj}^s, $(s = 1, 2, ..., t \, ; \, r = 1, 2, ... k \, ; \, j = 1, 2, ..., n)$, the number of votes obtained by the jth candidate in the rth position within the sth group of undesirable voters. Table 7.6 shows how these votes are aggregated.

Because the positive effect of lower priority votes for each category of voters is greater than the positive effect of higher priority votes in that category of voters; the negative effect of lower priority votes related to each category of voters in the category of undesirable voters will be greater than the negative effect of higher priority votes in that category. Therefore, if $\left(w_1^s, w_2^s, ..., w_k^s\right)$ and $\left(u_1^s, u_2^s, ..., u_k^s\right)$ are respectively the weight vectors of voting priorities in desirable and undesirable categories of voters, corresponding to category s $(s = 1, 2, ..., t)$, then we have $w_1^s > w_2^s > ... > w_k^s$, $w_r^1 > w_r^2 > ... > w_r^t, u_1^s > u_2^s > ... > u_k^s$ and $u_r^1 > u_r^2 > ... > u_r^t$ $(r = 1, 2, ..., k; s = 1, 2, ..., t)$. If the DEA policy is to be used in the management of this group voting system, the weight vectors should be chosen in such a way that the positive effect of the votes of desirable voters is the largest and the negative effect of the votes of undesirable voters is the least effective in evaluating the candidates. Model (7.9) was presented by Soltanifar et al. [9] to perform the group preferential voting process in the presence of desirable and undesirable voters for the assumed candidate p $(p = 1, 2, ..., n)$.

Table 7.6 Aggregation of votes of desirable and undesirable voters (group version)

Voting priorities	First category								...	rth category							
	Desirable				Undesirable				...	Desirable				Undesirable			
	C_1	C_2	...	C_n	C_1	C_2	...	C_n	...	C_1	C_2	...	C_n	C_1	C_2	...	C_n
1st	y^1_{11}	y^1_{12}	...	y^1_{1n}	x^1_{11}	x^1_{12}	...	x^1_{1n}	...	y^t_{11}	y^t_{12}	...	y^t_{1n}	x^t_{11}	x^t_{12}	...	x^t_{1n}
2nd	y^1_{21}	y^1_{22}	...	y^1_{2n}	x^1_{21}	x^1_{22}	...	x^1_{2n}	...	y^t_{21}	y^t_{22}	...	y^t_{2n}	x^t_{21}	x^t_{22}	...	x^t_{2n}
...	...	...	...	...	...	...	...	...	...	...	...	...	...	...	...	...	...
mth	y^1_{m1}	y^1_{m2}	...	y^1_{mn}	x^1_{m1}	x^1_{m2}	...	x^1_{mn}	...	y^t_{m1}	y^t_{m2}	...	y^t_{mn}	x^t_{m1}	x^t_{m2}	...	x^t_{mn}

$$Max\, Z_p^+ = \sum_{s=1}^{t} \sum_{r=1}^{k} w_r^s y_{rp}^s$$

$$Min\, Z_p^- = \sum_{s=1}^{t} \sum_{r=1}^{k} u_r^s x_{rp}^s$$

$$s.t.\ \sum_{s=1}^{t} \sum_{r=1}^{k} w_r^s y_{rj}^s \leq 1, \quad j = 1, 2, \ldots, n$$

$$\sum_{s=1}^{t} \sum_{r=1}^{k} u_r^s x_{rj}^s \geq 1, \quad j = 1, 2, \ldots, n \tag{7.9}$$

$$w_r^s - w_{r+1}^s \geq d^+(r, \varepsilon), \quad r = 1, 2, \ldots, k-1; \ s = 1, 2, .., t$$

$$w_k^s \geq d^+(k, \varepsilon), \quad s = 1, 2, .., t$$

$$w_r^s - w_r^{s+1} \geq \overline{d}^+(s, \varepsilon), \quad s = 1, 2, \ldots, t-1; \ r = 1, 2, .., k$$

$$w_r^t \geq \overline{d}^+(t, \varepsilon), \quad r = 1, 2, .., k$$

$$u_r^s - u_{r+1}^s \geq d^-(r, \varepsilon), \quad r = 1, 2, \ldots, k-1; \ s = 1, 2, .., t$$

$$u_k^s \geq d^-(k, \varepsilon), \quad s = 1, 2, .., t$$

$$u_r^s - u_r^{s+1} \geq \overline{d}^-(s, \varepsilon), \quad s = 1, 2, \ldots, t-1; \ r = 1, 2, .., k$$

$$u_r^t \geq \overline{d}^-(t, \varepsilon), \quad r = 1, 2, .., k$$

Model (7.9) is a bi-objective linear programming model, where $d^+(., \varepsilon)$, $\overline{d}^+(., \varepsilon)$, $d^-(., \varepsilon)$ and $\overline{d}^-(., \varepsilon)$ are the discrimination intensity functions related to desirable and undesirable category voters, corresponding to voting priorities and categories of voters. This model considers the strategy of DEA for desirable voters and undesirable voters simultaneously. It also makes it possible to determine the distance between voting priorities and categories of voters in both desirable and undesirable categories after interacting with the decision maker, and thus the final results are more reasoned and more acceptable to the decision maker. In model (7.9), if the distance between the voting priorities and categories of voters is the same for both desirable and undesirable voters, $d^+(., \varepsilon)$ and $d^-(., \varepsilon)$ and also $\overline{d}^+(., \varepsilon)$ and $\overline{d}^-(., \varepsilon)$ will be the same. Also the optimal value of both efficiency scores Z_p^+ and Z_p^- is limited to 1. This model is capable of being solved by methods which resolve multi-objective problems, such as, the conversion of objective function to constraints method, weighting method, absolute priority method, and the goal programming method and like them. In utilizing the goal programming method and since, the goal of first and second objective functions are equal to 1, model (7.9) is converted to model (7.10) which is a linear programming model.

Min $d_1 + d_2$

s.t. $\displaystyle\sum_{s=1}^{t}\sum_{r=1}^{k} w_r^s y_{rp}^s + d_1 = 1$

$\displaystyle\sum_{s=1}^{t}\sum_{r=1}^{k} u_r^s x_{rp}^s - d_2 = 1$

$\displaystyle\sum_{s=1}^{t}\sum_{r=1}^{k} w_r^s y_{rj}^s \le 1, \quad j = 1, 2, \ldots, n$

$\displaystyle\sum_{s=1}^{t}\sum_{r=1}^{k} u_r^s x_{rj}^s \ge 1, \quad j = 1, 2, \ldots, n$ (7.10)

$w_r^s - w_{r+1}^s \ge d^+(r, \varepsilon), \quad r = 1, 2, \ldots, k-1; \ s = 1, 2, \ldots, t$

$w_k^s \ge d^+(k, \varepsilon), \quad s = 1, 2, \ldots, t$

$w_r^s - w_r^{s+1} \ge \overline{d}^+(s, \varepsilon), \quad s = 1, 2, \ldots, t-1; \ r = 1, 2, \ldots, k$

$w_r^t \ge \overline{d}^+(t, \varepsilon), \quad r = 1, 2, \ldots, k$

$u_r^s - u_{r+1}^s \ge d^-(r, \varepsilon), \quad r = 1, 2, \ldots, k-1; \ s = 1, 2, \ldots, t$

$u_k^s \ge d^-(k, \varepsilon), \quad s = 1, 2, \ldots, t$

$u_r^s - u_r^{s+1} \ge \overline{d}^-(s, \varepsilon), \quad s = 1, 2, \ldots, t-1; \ r = 1, 2, \ldots, k$

$u_r^t \ge \overline{d}^-(t, \varepsilon), \quad r = 1, 2, \ldots, k$

$d_1, d_2 \ge 0$

In model (7.10), the purpose is to minimize the deviation variables of each objective function from the goal; where the votes of the voters are used in two categories, desirable and undesirable. Now, if $\left(\left(w_1^{1*}, \ldots, w_k^{1*}, u_1^{1*}, \ldots u_k^{1*}\right), \ldots, \left(w_1^{t*}, \ldots, w_k^{t*}, u_1^{t*}, \ldots u_k^{t*}\right)\right)$ is the optimal solution of the model (7.10), then Z_p^* obtained from the Eq. (7.11), can be the performance score of the hypothetical candidate p ($p = 1, 2, \ldots, n$) and be the basis for evaluating the candidates.

$$Z_p^* = \frac{Z_p^+}{Z_p^-} = \frac{\displaystyle\sum_{s=1}^{t}\sum_{r=1}^{k} w_r^{s*} y_{rp}^s}{\displaystyle\sum_{s=1}^{t}\sum_{r=1}^{k} u_r^{s*} x_{rp}^s} \qquad (7.11)$$

In this section, the issue of the presence of undesirable voters in the process of group voting has been raised theoretically and a model has been presented to deal with the process of group preferential voting in the presence of desirable and undesirable voters; but like what was presented in the previous section, the practical application of this model is still a question that needs to be answered. In the following, the application of the proposed model is presented.

As mentioned earlier, the classification of voters into desirable and undesirable categories may raise the question of whether this is not a violation of voting principles? What application can this modelling have in industry and society? These questions were answered in the preferential voting process by proposing a MADM method. In order to solve the ambiguities and answer the questions in the process of group preferential voting, this process is also introduced as an application in presenting a MADM method. In many MADM models, to choose an alternative from among a number of alternatives or to rank several alternatives considering several criteria, we are faced with criteria of various benefits and costs. These criteria are divided into categories that have different effects in the evaluation process. For example, in choosing a car, prestige and comfort are criteria of profit, and when it is higher, it is more desirable; but miles per gallon (MPG) and price are cost criteria and when they are less, they will be desirable for the decision maker. Also, prestige and comfort or MPG and price have different priorities in each category and are not equally effective in the result. In most of the models presented for these problems, cost criteria are converted into profit criteria in the normalization process, and then the ranking process takes place, and a predetermined weight is determined for each criterion to adjust its effectiveness. The proposed group preferential voting model in the presence of desirable and undesirable voters can solve the problem without changing by considering a benefit criterion as a desirable voter in each category of voters and a cost criterion as an undesirable voter in each category of voters. To solve the nature of each criterion and the need to weight the criteria. In fact, the results obtained by each alternative will be effective in each benefit criterion with a positive impact and in each cost criterion with a negative impact in the ranking of that alternative. Also, the positive (negative) impact of criteria with higher priority will be greater in the desirable (undesirable) category. Therefore, based on the scores of each alternative in each criterion, the alternatives can be ranked in that criterion. Placing any of the alternatives by any criteria in any voting priority is considered as a vote for that alternative in that priority. This vote can be desirable or undesirable depending on the criterion of profit or cost. Then votes are aggregated using model (7.10).

Let us describe the proposed method. Assume we want to evaluate n similar alternatives $(A_1, A_2, \ldots, A_n)$ by considering s multiple attributes $(C_1, C_2, \ldots, C_s)$. Also, assume that some criteria are profit type and some are cost type, and in each category the criteria are divided into sub-categories with different priorities. The overall decision matrix is also the same matrix presented in Eq. (7.4).

Step 1: Normalize the decision matrix (7.4) using Eq. (7.12).

$$\bar{t}_{pq} = \frac{t_{pq} - t_q^-}{t_q^+ - t_q^-}; \quad p = 1, 2, \ldots, n; \ q = 1, 2, \ldots, s. \tag{7.12}$$

where $t_q^- = \min_{1 \leq p \leq n} t_{pq}$ and $t_q^+ = \max_{1 \leq p \leq n} t_{pq}$, $q = 1, 2, \ldots, s$. This method is one of the many approaches suggested in the literature for normalizing decision matrices in MADM. This method is one of the many approaches suggested in the literature for normalizing decision matrices in MADM. It should be noted that normalization is used to standardize the unit of measurement regardless of the attribute type.

Step 2: Categorize the attributes using the judgments of the experts. This categorization describes the importance of attributes. Suppose the experts divide the attributes into t categories so that the importance of the characteristics embedded in category a is higher than that of the characteristics embedded in category b whenever $a < b$. Each attribute category can be divided into benefit and cost categories.

Step 3: Form the voting matrix shown in Table 7.6 using equations. (7.13) and (7.14). The number of candidates and voting priorities will be equal in the resulting Table 7.6 ($k = n$) since we need to identify and allocate the priority of each alternative among all the alternatives in each category.

$$y^s_{rj} = \sum_{\substack{q\in s\text{th category of profit attributes}\\ \bar{t}_{jq}\text{ is the }r\text{th priority in the column}}} \bar{t}_{jq} \tag{7.13}$$

$$x^s_{rj} = \sum_{\substack{q\in s\text{th category of cost attributes}\\ \bar{t}_{jq}\text{ is the }r\text{th priority in the column}}} \bar{t}_{jq} \tag{7.14}$$

Step 4: Solve model (7.10) and calculate the score of the alternatives using Eq. (7.11).

The flowchart of the proposed method is presented in Fig. 7.2.

In the following, to show the applicability of the MADM method, the example used by Soltanifar et al. [9] is presented.

Example 7.2. Saaty [10] introduced this numerical example to choose the best car among three alternatives (Acura TL, Toyota Camry, and Honda Civic) by considering different priorities for the following four attributes: Prestige, Comfort, Price, and MPG. This example is usually presented in the MADM literature in a hierarchical structure or a decision matrix. Table 7.7 shows the decision matrix for this problem. It is clear that the Comfort and Prestige attributes are of the benefit type, while the Price and MPG attributes are of the cost type. Experts have judged that Comfort and Price are more important than Prestige and MPG. After normalizing the decision matrix using Eq. (7.12), The (normalized) matrix presented in Table 7.8 is obtained. Then the voting matrix is formed using Eqs. (7.13) and (7.14) such as Table 7.9.

The results in Table 7.10 are obtained after solving model (7.10) for different discrimination intensity functions and calculating Eq. (7.11). Note that it is possible to interact with the decision maker so that the most suitable discrimination intensity functions are selected and finally a satisfactory solution is obtained. In this example, all discrimination intensity functions are assumed equal to ε and model (7.10) is solved for different ε. ε_{max} is the maximum value of epsilon for which model (7.10) remains feasible.

Thus, based on the values obtained in Table 7.10, the alternatives can be ranked. The results show that Acura TL $\succ$ Toyota Camry $\succ$ Honda Civic.

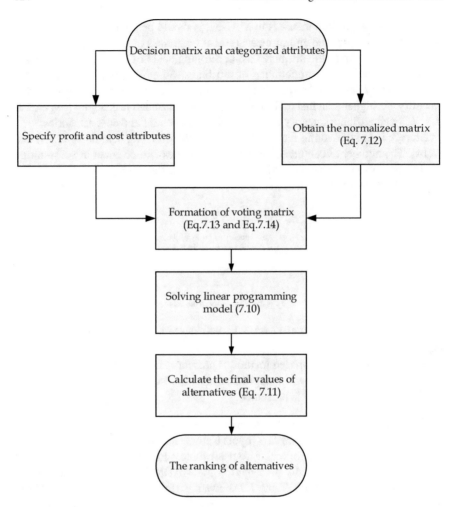

Fig. 7.2 The flowchart of the MADM method by group preferential voting process

Table 7.7 Decision matrix for choosing the best car

Decision matrix	First category		Second category	
	Comfort	Price	Prestige	MPG
Attribute type	Benefit	Cost	Benefit	Cost
Acura TL	0.705	0.937	0.707	0.818
Toyota Camry	0.211	0.806	0.07	0.5
Honda Civic	0.084	0.257	0.223	0.75

Table 7.8 Normalized decision matrix for choosing the best car

Normalized Decision Matrix	First Category		Second Category	
	Comfort	Price	Prestige	MPG
Attribute type	Benefit	Cost	Benefit	Cost
Acura TL	1.000	1.000	1.000	1.000
Toyota Camry	0.205	0.807	0.000	0.000
Honda Civic	0.000	0.000	0.240	0.786

Table 7.9 Voting matrix for choosing the best car

Voting Matrix	First category, desirable (Comfort)			First category, undesirable (Price)			Second category, desirable (Prestige)			Second category, undesirable (MPG)		
Voting Priority	1	2	3	1	2	3	1	2	3	1	2	3
Acura TL	1.000	0.000	0.000	1.000	0.000	0.000	1.000	0.000	0.000	1.000	0.000	0.000
Toyota Camry	0.000	0.205	0.000	0.000	0.807	0.000	0.000	0.000	0.000	0.000	0.000	0.000
Honda Civic	0.000	0.000	0.000	0.000	0.000	0.000	0.000	0.240	0.000	0.000	0.786	0.000

Table 7.10 The results of model (7.1) and Eq. (7.11) for different ε values

The results of model (7.1) and Eq. (7.11)	$\varepsilon_{max} = 0.143$	$\varepsilon = 0.071$	$\varepsilon = 0.048$	$\varepsilon = 0.010$	$\varepsilon = 0.007$	$\varepsilon = 0.004$
Acura TL	0.336	0.362	0.372	0.389	0.390	0.391
Toyota Camry	0.077	0.135	0.155	0.190	0.192	0.196
Honda Civic	0.069	0.094	0.103	0.117	0.118	0.119

7.4 Summary

In this chapter, the concept of undesirable voters was proposed as a theory and preferential voting models were presented to handle this type of voters. Undesirable voters are voters whose votes have a negative impact on the evaluation of candidates. After stating this theory, as an application, a method was presented that determines the weights of criteria and alternatives in MADM problems. Theory and application for preferential voting as well as group preferential voting were presented. In the presented methods for solving MADM problems, there is no need to change the nature of the criteria in the normalization process, and through the discrimination intensity functions, it is possible to interact more with the decision maker during the problem solving.

Appendix

In this section, GAMS software codes for the models described in this chapter are provided.

Model (7.2):

```
Sets
    R                  /O1 * O4/
    J                  /DMU1*DMU4/
    k(R),K1(R);

Alias (J,L);

Parameters
    b(j),EPI,EPIP,EF,EFP,ZP,YO(R),YPO(R);


Table Y(J,R)
          O1       O2        O3        O4
DMU1    0        0.041699  0.209517  0
DMU2    0.047    0         0.209517  0
DMU3    0.369    0         0.035338  0
DMU4    0        0.312712  0         0.028977;

Table YP(J,R)
          O1       O2        O3        O4
DMU1    0.819    1.313371  0         0
DMU2    0        0.80643   0.6496    0.488189
DMU3    1.597    0         0         0.781289
DMU4    0        0         1.30013   0.6148 ;

Variables
      z ,U(R),V(R),D1,D2;
    positive variables
      U,V,D1,D2;

Equation
      Objective,Const1,  Const2,const3,const4,const5,const6,CONST7,CONST8;

Objective..          z=e=D1+D2;
Const1(J)..          Sum(R,U(R)*YO(R))+D1    =E=        1;
Const2(J)..          Sum(R,V(R)*YPO(R))-D2   =E=        1;
Const3(J)..          Sum(R,U(R)*Y(J,R))      =L=        1;
Const4(J)..          Sum(R,V(R)*YP(J,R))     =G=        1;
Const5(R)$K(R)..     U(R)-U(R+1)             =G=        EPI;
CONST6..             U('O4')                 =G=        EPI;
Const7(R)$K1(R)..    V(R)-V(R+1)             =G=        EPI;
CONST8..             V('O4')                 =G=        EPI;

Model                VOTING_Model                      /Objective,Const1,
Const2,const3,const4,const5,const6,CONST7,CONST8/;

File  Super_CCR_I /.txt/;

Puttl Super_CCR_I 'Title ' System.title, @60 'Page ' System.page//;

Put Super_CCR_I;

Put @10'E', @22'EP', @34'ZP'/;
```

```
EPI   = 0.0001;

Put 'EPI= 'EPI:12:8/;
loop(R,
         K(R)=Yes;
         K1(R)=YES;
);

K('O4')=NO;
K1('O4')=NO;

LOOP(L,
         LOOP(R,
                 YO(R)=Y(L,R);
                 YPO(R)=YP(L,R);
         );

         Solve VOTING_Model using LP MINimizing z ;


         EF= Sum(R,U.L(R)*Y(L,R)) ;
         EFP= Sum(R,V.L(R)*YP(L,R));
         ZP=EF/EFP;

         PUT L.TL:8;
         PUT EF:12:9;
         PUT EFP:12:9;
         PUT ZP:12:9;
         Put/;
);
```

Model (7.10):

```
Sets
   i "Iputs"      /i1 * I2/
   r "Outputs"    /1 * 3/
   j "Units"      /DMU1*DMU3/
   K(R),K1(R)
;

Alias(L,J);


Parameters
X(I,R,J),Y(I,R,J),XP(I,R),YP(I,R),EPIP,EF
;
```

TABLE Y1(J,R)

	1	2	3
DMU1	0.951832268	0	0
DMU2	0	0.284874622	0
DMU3	0	0	0.113409802;

```
TABLE Y2(J,R)
           1                  2                  3
DMU1    0.949461271        0                  0
DMU2    0                  0.299476469        0
DMU3    0                  0                  0.094006066

;




TABLE X1(J,R)
           1                  2                  3
DMU1    0.742236953        0                  0
DMU2    0                  0.638466365        0
DMU3    0                  0                  0.203580466
;

TABLE X2(J,R)
           1                  2                  3
DMU1    0.690184042        0                  0
DMU2    0                  0.613403176        0
DMU3    0                  0                  0.383904326
;




LOOP(J,
        LOOP(R,
                X('I1',R,J)=X1(J,R);
                X('I2',R,J)=X2(J,R);

                Y('I1',R,J)=Y1(J,R);
                Y('I2',R,J)=Y2(J,R);
        );
);




Variables
D1,D2,Z,W;
     Positive Variables
                   D1,D2,U(I,R),V(I,R),EPI(I),EEP;

Equations
       Objective ,Const1 ,Const2,CONST3 ,Const4,CONST5,CONST6 ,CONST7
,CONST8
CONST9,CONST10
ObjectiveA ,Const1A ,Const2A,CONST3A ,Const4A,CONST5A,CONST6A ,CONST7A
,CONST8A
CONST9A,CONST10A
;
```

```
Objective..        Z =E=EEP;
Const1..           SUM(I,SUM(R,U(I,R)* YP(I,R))) + D1 =E= 1;
Const2..           SUM(I,SUM(R,V(I,R)* XP(I,R))) - D2 =E= 1;

Const3(J)..        SUM(I,SUM(R,U(I,R)* Y(I,R,J)))  =L= 1;
Const4(J)..        SUM(I,SUM(R,V(I,R)* X(I,R,J)))  =G= 1;

CONST5(I,R)$K(R)..      U(I,R)-U(I,R+1) =G= EEP;
CONST6(I)..            U(I,'3') =G= EEP;

CONST7(I,R)$K(R)..      V(I,R)-V(I,R+1) =G= EEP;
CONST8(I)..            V(I,'3') =G= EEP;

CONST9(I,R) ..            U(I,R)-U(I+1,R)=G=EEP;
CONST10(I,R)..            V(I,R)-V(I+1,R)=G=EEP;


ObjectiveA..       Z =E=D1+D2 ;
Const1A..          SUM(I,SUM(R,U(I,R)* YP(I,R))) + D1 =E= 1;
Const2A..          SUM(I,SUM(R,V(I,R)* XP(I,R))) - D2 =E= 1;

Const3A(J)..       SUM(I,SUM(R,U(I,R)* Y(I,R,J)))  =L= 1;
Const4A(J)..       SUM(I,SUM(R,V(I,R)* X(I,R,J)))  =G= 1;

CONST5A(I,R)$K(R)..     U(I,R)-U(I,R+1) =G= EPIP;
CONST6A(I)..           U(I,'3') =G= EPIP;

CONST7A(I,R)$K(R)..     V(I,R)-V(I,R+1) =G= EPIP;
CONST8A(I)..           V(I,'3') =G= EPIP;

CONST9A(I,R) ..           U(I,R)-U(I+1,R)=G= EPIP;
CONST10A(I,R)..           V(I,R)-V(I+1,R)=G= EPIP;



Model   MODEL_1  /Objective  ,Const1  ,Const2,CONST3  ,Const4,CONST5,CONST6
,CONST7 ,CONST8
CONST9,CONST10/;
Model       MODEL_2      /ObjectiveA      ,Const1A      ,Const2A,CONST3A
,Const4A,CONST5A,CONST6A ,CONST7A ,CONST8A
CONST9A,CONST10A/;

File CCR_AR /Results.txt/;

Puttl CCR_AR  'Title ' System.title, @60'Page ' System.page//;

Put CCR_AR;

Put @7'Efficiency', @21'Input-weights', @40'Output-weights'/;


LOOP(R,K(R)=YES);
```

```
K('3')=NO;


Loop(L,
     LOOP(I,
         Loop(R,XP(I,R)=X(I,R,L));
         Loop(R,YP(I,R)=Y(I,R,L));
          );
                        Solve MODEL_1 Using LP MAXimizing z;
                        EPIP=EEP.L;

                        Solve MODEL_2 Using LP MINimizing z;

                        Put L.TL:6;

                        PUT EPIP:12:10' ';
                        EF=(SUM(I,SUM(R,U.L(I,R)*
YP(I,R))))/(SUM(I,SUM(R,V.L(I,R)* XP(I,R))));
                        PUT EF:12:6' ';


                        Solve MODEL_2 Using LP MINimizing z;
                        EPIP=EEP.L/2;
                        PUT EPIP:12:10' ';
                        EF=(SUM(I,SUM(R,U.L(I,R)*
YP(I,R))))/(SUM(I,SUM(R,V.L(I,R)* XP(I,R))));
                        PUT EF:12:6' ';
                     PUT/;

  );
```

References

1. Cook, W., Kress, M.: A data envelopment model for aggregating preference rankings. Manage. Sci. **36**, 1302–1310 (1990)
2. Färe, R., Grosskopf, S., Lovell, C.A.K., Pasurka, C.: Multilateral productivity comparisons when some outputs are undesirable: a nonparametric approach. Rev. Econ. Stat. **71**, 90–98 (1989)
3. Scheel, H.: Undesirable outputs in efficiency valuations. Eur. J. Oper. Res. **132**(2), 400–410 (2001)
4. Hua, Z., Bian, Y.: DEA with undesirable factors. In: Modeling Data Irregularities and Structural Complexities in Data Envelopment Analysis. MA, Springer, Boston (2007)
5. Halkos, G., Petrou, K.N.: Treating undesirable outputs in DEA: a critical review. Econ. Anal. Pol. **62**, 97–104 (2019)
6. Zhu, W., Xu, M., Cheng, C.P.: Dealing with undesirable outputs in DEA: An aggregation method for a common set of weights. J. Operat. Res. Soc. **71**(4), 579–588 (2020)
7. Soltanifar, M., Sharafi, H.: Preferential voting in the presence of undesirable voters. Inter. J. Appl. Decis. Sci. (2022)

8. Soltanifar, M., Heidariyeh, S.A.: Employee performance evaluation using a new preferential voting process. Inno. Manage. Operat. Strat. **1**(3), 202–220 (2020)
9. Soltanifar, M., Tavana, M., Santos-Arteaga, F.J., Sharafi, H.: A hybrid multi-attribute decision-making and data envelopment analysis method for solving problems with heterogeneous attributes. Environmental Science & Policy (vol. submitted, 2022)
10. Saaty, T.L.: The modern science of multi attributes decision making and its practical applications: the AHP/ANP approach. Oper. Res. **61**(5), 1101–1118 (2013)

Chapter 8
Hybrid Multi-attribute Decision-Making Methods Based on Preferential Voting

Abstract In addition to the direct application of preferential voting models in a voting process, these models can be combined with other Multi-Attribute Decision Making (MADM) methods to improve them and achieve new hybrid models. In this chapter, some hybrid MADM methods based on preferential voting models are presented. New hybrid models with the help of preferential voting models have become more powerful tools for decision support.

8.1 Introduction

Decision-making includes the correct expression of targets, determining different solutions, evaluating their feasibility, evaluating the consequences and results of implementing each of the solutions, and finally choosing and implementing the selected solution. The quality of management basically depends on the quality of decision-making, because the quality of plans and programs, the effectiveness and efficiency of strategies, and the quality of the results obtained from their application are all dependent on the quality of the decisions made by the manager. In most cases, decisions are desirable and satisfactory to the decision-maker when the decision-making process is based on several criteria. Criteria may be quantitative or qualitative. In the Multi-Criteria Decision-Making (MCDM) methods that have attracted the attention of researchers in recent decades, several measurement criteria are used instead of one optimality measurement criterion. MCDM methods are divided into two major categories: Multi-Objective Decision-Making (MODM) and Multi-Attribute Decision-Making (MADM). In general, MODM are used to design and MADM are used to select the best alternative. The main difference between MODM and MADM is that the first one is defined in the continuous decision-making space and the second one is defined in the discrete decision-making space. The methods of solving MCDM problems are repeatedly examined by researchers and their strengths and weaknesses are identified. Also, these studies have led to the design of several hybrid methods to solve some weaknesses. Preferential voting models are one of the tools that have provided significant results in combination with

© The Author(s), under exclusive license to Springer Nature Switzerland AG 2023
M. Soltanifar et al., *Preferential Voting and Applications: Approaches Based on Data Envelopment Analysis*, Studies in Systems, Decision and Control 471, https://doi.org/10.1007/978-3-031-30403-3_8

MCDM methods. Soltanifar [1] improved some well-known MODM methods using the concepts of discrimination intensity functions and provided a complete analysis of the proposed hybrid methods. Because MADM is more applicable to real-world problems [2], it has been more widely developed by researchers than MODM in the last 60 years. This chapter, however, focuses on the application of preference voting models in improving the performance of MADM methods. Many presented hybrid methods are analyzed with the help of preferential voting models and the strengths of hybrid methods are expressed.

8.2 Preferential Voting and Methods Based on Pairwise Comparisons

Many MADM methods are based on pairwise comparisons between attributes or alternatives. In these methods, the weights of the attributes or the local weights of the alternatives are calculated after performing their pairwise comparisons, and the overall weights of the alternatives are presented for ranking based on this. In this section, these methods are briefly presented and some of their shortcomings are reported, then using the concept of preferential voting, hybrid methods will be presented to solve the mentioned shortcomings. All methods are implemented on a well-known example from the MADM literature.

8.2.1 Voting Analytical Hierarchy Process (VAHP)

Analytical Hierarchy Process (AHP) is one of the most popular MADM methods proposed by Saaty [3]. AHP is a structured technique for organizing and solving complex decision-making problems based on mathematics and psychology. AHP provides a comprehensive and logical framework for quantifying each element in a hierarchical structure. AHP starts with the selection of decision criteria and follows the principles of reciprocal condition, homogeneity, dependency, and expectations to prioritize each criterion. Saaty [3] highlights the following as the main advantages of AHP: unity in providing a model for problem solving, analytical and systematic approach in solving complex problems, problem solving power when criteria are interrelated, compliance with hierarchical structure in decision making, measurement Intangible and qualitative items, examining the compatibility of priorities, combining the usefulness of alternatives, maintaining balance in priorities, the possibility of group judgment and the possibility of improvement through repetition. The algorithm of this method is as follows.

Step 1. Create a hierarchical structure. First, the main criteria and alternatives that determine the decision-making problem are determined, and then the problem is divided into goal levels, criteria, sub-criteria and alternatives. Each element of this hierarchy depends on its higher level element and this dependency continues linearly up to the highest level. In addition, the evaluation process must be repeated whenever there is a change in the hierarchical structure.

Step 2. Form a pairwise comparison matrix. The elements of each level are pairwise compared, which leads to the formation of pairwise comparison matrices. A 9-point scale is used to determine importance and priority in pairwise comparisons. The preferences at this stage must meet the conditions of reciprocal and homogeneity.

Step 3. Calculation of inconsistency rate. Considering that the judgment of experts may lead to the formation of inconsistent pairwise comparison matrices, an empirical rate is proposed to evaluate their consistency and hierarchical structure. In case of disagreement, the results are returned to the experts for re-examination. The algorithm for calculating the mismatch rate of a pairwise comparison matrix (D) is expressed as follows:

a. Calculate the weighted sum vector. Calculate the overall priority vector by multiplying the pairwise comparison matrix (D) by the local priority vector.
b. Calculate the consistency vector. Calculate the consistency vector by dividing the elements of the overall priority vector coordinate-wise by those of the local priority vector. That is, each element of the consistency vector is obtained by dividing the corresponding element of the weighted sum vector by that of the local priority vector. The components of the consistency vector are actually λ_{max} estimates.
c. Calculate the largest Eigenvalue of the pairwise comparison matrix (λ_{max}). The average of the elements of the consistency vector is equal to λ_{max}.
d. Calculate the inconsistency index (II). Assuming that the pairwise comparison matrix (D) is an $m \times m$ matrix, the inconsistency index equals $\frac{\lambda_{max}-m}{m-1}$.
e. Define the inconsistency ratio (IR). The IR is given by $\frac{II}{IRI}$, where IRI is the inconsistency random index, whose value is extracted from Table 8.1. The values in this table are determined via simulation.

Satty [3] suggested that if the inconsistency ratio is less than or equal to 0.1, the results of the pairwise comparisons are acceptable. Otherwise, they are returned to the expert(s) for review and reconsideration.

Step 4. Calculate local priorities. The local priorities of the criteria and the alternatives relative to each criterion are obtained using different weighting methods. The most common weighting methods include the sum of rows, columns, arithmetic mean, geometric mean, eigenvector, ordinary least squares, and logarithmic least squares.

Table 8.1 Inconsistency index of random matrix (I.I.R.)

M	1	2	3	4	5	6	7	8	9	10	11	12	13	14	15
I.I.R	0	0	0.58	0.90	1.12	1.24	1,32	1.41	1.45	1.45	1.51	1.52	1.56	1.57	1.59

Step 5. Calculate the overall priority of the alternatives. The overall priority of each alternative is equal to the sum of the product of the local priority of the alternative relative to each weighted criterion.

Step 6. Rank the alternatives. The alternatives are ranked based on their overall priorities. The higher the overall priority of an alternative, the better its ranking position.

One of the most important issues in the AHP method is the necessity of precision in the formation of pairwise comparison matrices by experts, because lack of precision in this regard makes the final results unusable. Although the calculation of the inconsistency rate and the indicator for the acceptance of the matrices can assure the decision maker of the correctness of the results to some extent, but this complex unconscious process makes the participants reluctant and unmotivated, and this tedious process itself causes a decrease accuracy is the final result. Therefore, in general, despite the many advantages, this method has a significant defect in the ranking of options when there are many elements of each level.

Preferential voting models can be used as a tool to solve this problem. Liu and Hai [4] presented the VAHP hybrid method by using this tool. This method was developed and used by [5–9]. The algorithm of this method is as follows.

Step 1. Create a hierarchical tree.

Step 2. Determine the priority of criteria and that of the alternatives per criterion. The elements composing each level are compared to other related elements located at a higher level, and their priorities are determined.

Step 3. Calculate the local priorities: After solving model (4.14) and obtaining the optimal Z_j^*, $j = 1, 2, ..., m$, values, the relative local priority weights of the criteria and alternatives are computed using Eq. (8.1).

$$ w_j = \frac{Z_j^*}{\sum_{k=1}^m Z_k^*}, \quad j = 1, 2, ..., m \tag{8.1} $$

Step 4. Calculate the overall priority of the alternatives.

Step 5. Rank the alternatives.

Compared to the AHP method, the VAHP method does not require the formation of paired comparison matrices, and this leads to the elimination of the possibility of inconsistency of judgments. Also, this method has more flexibility in the structure of group decision making [10]. Also, this method has a better performance in increasing the motivation of experts to participate in the ranking process.

8.2.2 Best–Worst VAHP Method (BWVAHP)

One of the methods of weighting attributes and alternatives is the Best–Worst Method (BWM). BWM was presented by Rezaei [11] and many studies were conducted on it [12–17]. In this method, after determining the best and worst element, a pairwise comparison between other elements and the best and worst found becomes the basis for defining a mathematical programming model. In this method, there is a possibility of inconsistency in judgments, and therefore a formula for calculating the inconsistency rate has been provided to confirm the correctness of the comparisons. The BWM algorithm is as follows.

Step 1. Identify the influential elements for the purpose of the problem by interacting with the decision-maker(s), $E_1, E_2, ..., E_m$.

Step 2. Determine the best (E_B) and the worst (E_W) criteria and sub-criteria among those selected based on the opinion of the decision-maker.

Step 3. After interacting with the decision-maker, determine the preferences of the best criterion over the other criteria based on a 9-point scale, $(a_{Bj}, \ j = 1, 2, ..., m)$.

Step 4. After interacting with the decision-maker, determine the preferences of the other criteria over the worst criterion using a 9-point scale, $(a_{jW}, \ j = 1, 2, ..., m)$.

Step 5. Obtain the weights of the criteria by solving the mathematical programming model (8.2).

$$
\begin{aligned}
&\min \xi \\
&s.t. \\
&\left| w_B - a_{Bj} w_j \right| \leq \xi w_j, \quad j = 1, 2, ..., m \\
&\left| w_j - a_{jw} w_w \right| \leq \xi w_w, \quad j = 1, 2, ..., m \\
&m \sum_{j=1} w_j = 1 \\
&w_j \geq 0, \quad j = 1, 2, ..., m
\end{aligned}
\tag{8.2}
$$

Denote by $(w_1^*, w_2^*, \ldots, w_m^*)$ the optimal weights of the criteria obtained when solving model (8.2). The results obtained are validated through the inconsistency ratio of the system $C.R. = \frac{\xi^*}{C.I.}$, where ξ^* is the optimal value of the objective and $C.I.$ is extracted from Table 8.2. If the inconsistency ratio is close to zero, then it is plausible to rely on the judgments of the experts.

Although this method is presented as an element weighting method based on pairwise comparisons, this method can also be used as a MADM method. Tavana et al. [7] have done so by presenting a hybrid BWM-AHP method. By reducing the number of pairwise comparisons, the BWM-AHP method has provided an effective step in reducing the computational complexity and increasing the motivation of experts to

Table 8.2 Consistency index (*C.I.*)

m	1	2	3	4	5	6	7	8	9
(C.I.)	0	0.44	1.00	1.63	2.30	3.00	3.73	4.47	5.23

participate in the method process, but there is still a possibility of inconsistency in judgments. Using the concept of preferential voting, Tavana et al. [7] succeeded in presenting a hybrid method that, while eliminating many shortcomings, has provided the possibility of more interaction with the decision maker. The algorithm of this method is as follows.

Step 1. Create a hierarchical tree.

Step 2. Determine the priority of each level. The elements of each level are compared to other related elements located at a higher level, and their priorities are defined from best to worst. The presentation can be simplified by assuming that the priority order matches the index of each element.

Step 3. Determine the preferences of the best element per level. After interacting with the decision-maker, determine the preferences of the best criterion (the best alternative per criterion) relative to the other criteria (alternatives at the level selected) based on a 9-point scale. Without loss of generality, assume that the best element is the first element of the level (a_{j1}, $j = 1, 2, ..., m$).

Step 4. Determine the preferences of the worst element per level. After interacting with the decision-maker, determine the preferences of the other criteria (alternatives at the level selected) to the worst criterion (worst alternative per criterion) based on a 9-point scale. Without loss of generality, assume that the worst element is the last element of the level (a_{jm}, $j = 1, 2, ..., m$).

Step 5. Calculate the local priorities. The criteria and alternatives per criterion are obtained by solving the model (8.3). The discrimination intensity function $d(., \varepsilon)$, which determines the difference between the weight of each element and the immediately better one.

$$
\begin{aligned}
&\min \ \xi \\
&s.t. \\
&\left| w_1 - a_{1j} w_j \right| \le \xi w_j, \quad j = 1, 2, ..., m \\
&\left| w_j - a_{jm} w_m \right| \le \xi w_m, \quad j = 1, 2, ..., m \\
&\sum_{j=1}^{m} w_j = 1 \\
&w_j - w_{j+1} \ge d(j, \varepsilon), \quad j = 1, 2, ..., m - 1 \\
&w_m \ge d(m, \varepsilon)
\end{aligned}
\tag{8.3}
$$

Step 6. Calculate the inconsistency ratio of the judgments per level and the hierarchical inconsistency ratio.

Step 7. Calculate the overall priority of the alternatives.

Step 8. Rank the alternatives.

Despite the reduction of calculations due to the reduction of partial comparisons, this method allows for more interaction with the decision maker through discrimination intensity functions.

8.2.3 SWARA VAHP Method

Another weighting method that was used to solve the shortcomings of the AHP method is the Stepwise Weight Assessment Ratio Analysis (SWARA) method [18]. Using this method, Tavana et al. [7] presented the SWARA-AHP method, whose algorithm is as follows.

Step 1. Create a hierarchical tree.

Step 2. Prioritize criteria and alternatives over each criterion. The elements composing each level are compared to other related elements located at a higher level, and their priorities are determined. Criteria and alternatives are reevaluated after this step so that the criteria or alternatives with the smaller index values are given a higher priority.

Step 3. Determine the comparative importance of the average value. At each level, starting from the first element of the redesigned criteria or alternatives and after interacting with the decision-maker(s), the relative difference between each criterion or alternative and the previous one must be determined. The resulting variables are denoted s_j, $(j = 1, 2, ..., m)$, and known as the comparative importance of the average value. Note that $s_1 = 0$ for each level.

Step 4. Calculate the Local Priority: Eq. (8.4) is applied to compute the local priorities of the criteria and alternatives for each criterion while noting that $q_j = \frac{1}{\prod_{k=1}^{j}(1+s_k)}$, $(j = 1, 2, ..., m)$.

$$w_j = \frac{q_j}{\sum_{k=1}^{m} q_k} \tag{8.4}$$

Step 5. Calculate the overall priority of the alternatives.

Step 6. Rank the alternatives.

In the presented hybrid method, there is no need to form pairwise comparison matrices, and this solves the problem of inconsistency in judgments. Tavana et al.

[7] provided the possibility of more interaction with the decision maker by using the concepts of preferential voting and presented the hybrid SWARA-VAHP method. The algorithm of this method is as follows.

Step 1. Create a hierarchical tree.

Step 2. Determine the priority of criteria and alternatives per criterion. The elements composing each level are compared to other related elements located at a higher level, and their priorities are determined. The criteria and alternatives are re-indexed after this step so that the criterion or alternative with the smaller index is assigned a better priority.

Step 3. Determine the minimum comparative importance of the average value. At each level, starting from the first element of the redesigned criteria or alternatives and after interacting with the decision-maker(s), the relative difference between each criterion or alternative and the previous one must be determined.

Step 4. Calculate the local priorities. Equation (8.1) is applied to obtain the local priority of the criteria and alternatives per criterion, where Z_j^*, $(j = 1, 2, ..., m)$ the optimal values are obtained from model (8.5).

$$Z_p = \max \sum_{r=1}^{m} q_r v_{rp}$$

$s.t.$

$$\sum_{r=1}^{k} q_r v_{rj} \leq 1, \qquad j = 1, 2, ..., m \tag{8.5}$$

$$q_r \geq q_{r+1}(1 + s_{r+1}), \qquad r = 1, 2, ..., m - 1$$

$$q_m \geq \varepsilon_{\max}$$

Note that if the comparative importance of the average value of priority j to priority $(j - 1)$ is s_j, $(j = 2, ..., m)$ then, we must have $\frac{q_j}{q_{j+1}} \geq 1 + s_{j+1}$, $(j = 1, 2, ..., m - 1)$.

Step 5: Calculate the overall priority of the alternatives.

Step 6: Rank the alternatives.

Where $\varepsilon_{\max}$ is the maximum value for which the model (8.5) is feasible. In this way, while solving the problem of incompatibility in AHP, it is possible to interact more with the decision maker.

Example 8.1 In this section, we apply the hybrid methods on a familiar numerical example from the MADM literature [19]. Saaty [20] introduced this numerical example for choosing the best car among three alternatives (Acura TL (A.TL.), Toyota Camry (T.C.) and Honda Civic (H.C.)) by considering the following criteria: Prestige, Comfort, Price and Miles per gallon (MPG). The hierarchical tree of this

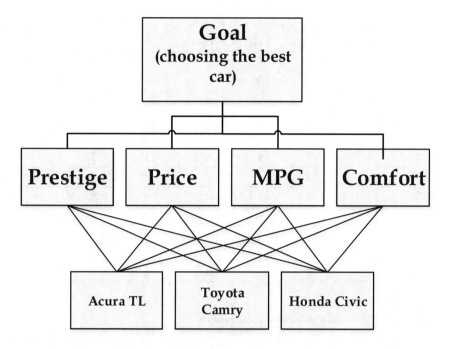

Fig. 8.1 Results of choosing the best car by GAHP method [19]

problem is presented in Fig. 8.1 and the related pairwise comparison matrices in Table 8.3.

The solution of this problem with the AHP method is given in Table 8.4. In this table, the local weights of the criteria and alternatives for each criterion, as well as the overall weights of the alternatives, are known.

AHP requires performing pairwise comparisons at each level. BWM-AHP incorporates pairwise comparisons between the best and worst elements per level and the other elements within the same level. Thus, for instance, to calculate the local priorities of the criteria, "price" being the best criterion and "prestige" being the worst one, it suffices to obtain pairwise comparisons of the best criterion compared to the other criteria, and the other criteria compared to the worst one.

BWM-VAHP requires the same information as BWM-AHP and prioritizes the elements composing each level, as was the case with VAHP. In addition, BWM-VAHP is sensitive to the type of discrimination intensity function defined in model (8.3), and the decision-maker must be careful in determining it. For instance, if all the discrimination intensity functions in model (8.13) are equal to zero, the same results as those of BWM-AHP would be obtained.

AHP, BWM-AHP retrieve information from the experts in the form of pairwise comparison matrices. However, in SWARA-AHP, the information received from the experts will be different, first prioritizing the elements of each level and then

Table 8.3 Pairwise comparison matrices for choosing the best car [19]

Goal	Prestige	Price	MPG	Comfort			
Prestige	1	$\frac{1}{4}$	$\frac{1}{3}$	$\frac{1}{2}$			
Price	4	1	3	$\frac{3}{2}$			
MPG	3	$\frac{1}{3}$	1	$\frac{1}{3}$			
Comfort	2	$\frac{2}{3}$	3	1			
Prestige	A.TL.	T.C.	H.C.	Price	A.TL.	T.C.	H.C.
A.TL.	1	8	4	A.TL.	1	$\frac{1}{4}$	$\frac{1}{9}$
T.C.	$\frac{1}{8}$	1	$\frac{1}{4}$	T.C.	4	1	$\frac{1}{5}$
H.C.	$\frac{1}{4}$	4	1	H.C.	9	5	1
MPG	A.TL.	T.C.	H.C.	Comfort	A.TL.	T.C.	H.C.
A.TL.	1	$\frac{2}{3}$	$\frac{1}{3}$	A.TL.	1	4	7
T.C.	$\frac{3}{2}$	1	$\frac{1}{2}$	T.C.	$\frac{1}{4}$	1	3
H .C.	3	2	1	H.C.	$\frac{1}{7}$	$\frac{1}{3}$	1

Table 8.4 Synthesis of the priorities of the alternatives

Goal (Buy best car)	Prestige	Price	MPG	Comfort	Synthesis of overall priorities
Priorities	0.099	0.425	0.169	0.308	
Acura TL	0.707	0.063	0.182	0.705	0.342
Toyota Camry	0.070	0.194	0.273	0.211	0.204
Honda Civic	0.223	0.743	0.545	0.084	0.454

determining the relative difference between each element and the previous, i.e., better one. Nevertheless, the structure of the information received and analyzed by SWARA-VAHP is the same as SWARA-AHP. To provide additional intuition, Tables 8.5 and 8.6 present a detailed implementation of SWARA-VAHP when ranking the set of alternatives based on the judgments of a team of experts.

A complete comparison of the results of the implementation of the mentioned methods on this example is available in Fig. 8.2.

8.3 Voting TOPSIS Method

Technique for Order Preference by Similarity to Ideal Solution (TOPSIS) is an MADM method for evaluating and prioritizing alternatives based on attributes according to their distance from positive and negative ideals. This method was proposed by Hwang and Yoon [21] and very soon it was proposed as one of the most used decision making methods. In this method, by defining positive ideal alternative (the best possible alternative) and negative ideal alternative (the worst possible

Table 8.5 Detailed implementation of SWARA-VAHP

Level	Goal (buy best car)	Expert 1	Expert 2	Expert 3	Expert 4	Expert 5	s1	s2	s3	s4	s5	Geometric mean	Local priorities	Normalized
Goal	Prestige	4	4	4	3	3	0	0	0	0	0	0	0.5432	0.1769
	Price	1	2	1	1	2	0.2	0.3	0.2	0.1	0.2	0.1888	1	0.3256
	MPG	3	3	3	4	4	0.5	0.4	0.5	0.6	0.5	0.4959	0.5623	0.1831
	Comfort	2	1	2	2	1	0.2	0.2	0.3	0.1	0.2	0.1888	0.9661	0.3145
Prestige	Acura TL	1	1	1	1	1	0	0	0	0	0	0	1	0.6972
	Toyota Camry	3	3	2	3	2	2	2	1	3	2	1.8882	0.2430	0.1694
	Honda Civic	2	2	3	2	3	3	3	4	2	3	2.9302	0.1914	0.1334
Price	Acura TL	3	3	2	3	2	0	0	0	0	0	0	0.1413	0.1075
	Toyota Camry	2	2	3	2	3	3.5	3.5	3	4	2.5	3.2588	0.1725	0.1313
	Honda Civic	1	1	1	1	1	2	2	2.5	1.5	2	1.9744	1	0.7612
MPG	Acura TL	3	3	2	3	2	0	0	0	0	0	0	0.3618	0.1907
	Toyota Camry	2	1	3	2	3	1	1	1.5	2	0.5	1.0845	0.5358	0.2824
	Honda Civic	1	2	1	1	1	1.5	1.5	1	0.5	2	1.1761	1	0.5270
Comfort	Acura TL	1	1	1	1	1	0	0	0	0	0	0	1	0.6960
	Toyota Camry	2	3	2	3	2	2	2	2.5	1.5	1.5	1.8640	0.2446	0.1702
	Honda Civic	3	2	3	2	3	3	3	2.5	3	3.5	2.9831	0.1923	0.1338

Table 8.6 Ranking results obtained from SWARA-VAHP

Goal (buy the best car)	Synthesis of overall priorities
Acura TL	0.4121
Toyota Camry	0.1779
Honda Civic	0.4100

alternative) and assuming a uniform increase or decrease in the desirability of each attribute, proximity to the positive ideal alternative and distance from the negative alternative became the basis for decision making. Also, Euclid's definition of distance was based on this method. This method has been used by researchers in many versions

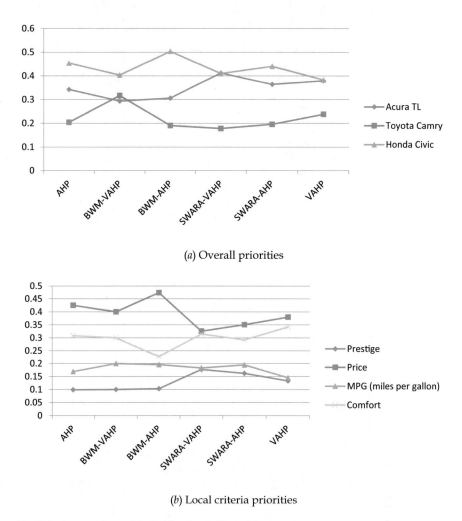

(*a*) Overall priorities

(*b*) Local criteria priorities

Fig. 8.2 A comparison of the local and overall priorities

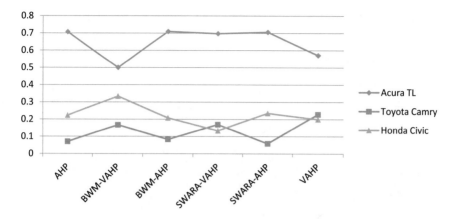

(c) Local alternative priorities according to prestige

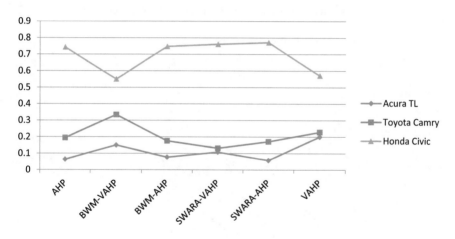

(d) Local alternative priorities according to price

Fig. 8.2 (continued)

and in different applications [22–25] and [26]. The algorithm of this method is as follows:

Suppose that n homogeneous alternatives $(A_1, A_2, ..., A_n)$ are to be ranked considering m attributes $(C_1, C_2, ..., C_m)$. The decision matrix of such a problem is as Eq. (8.6).

$$D = \begin{bmatrix} r_{11} & \cdots & r_{1m} \\ \vdots & \ddots & \vdots \\ r_{n1} & \cdots & r_{nm} \end{bmatrix} \tag{8.6}$$

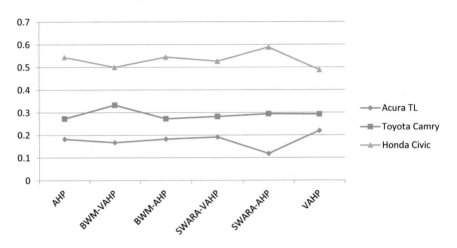

(*e*) Local alternative priorities according to MPG

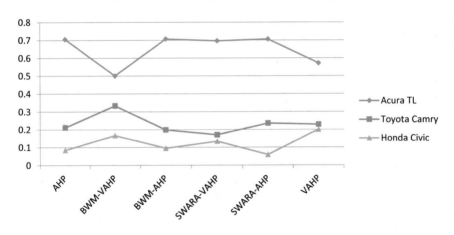

(*f*) Local alternative priorities according to comfort

Fig. 8.2 (continued)

Step 1. Normalize matrix D using normalization (8.7) called Euclidean normalization.

$$N_D = [n_{ij}]_{n \times m} \& n_{ij} = \frac{r_{ij}}{\sqrt{\sum_{k=1}^{n} r_{kj}^2}} \tag{8.7}$$

Step 2. Let's assume that the weights of the attributes are in the form of w_1, w_2, ..., w_m, which is specified by DM or calculated from methods such as Shannon's entropy method. Calculate the weighted normalized decision matrix from Eq. (8.8).

$$V = N_D \times \begin{bmatrix} w_1 & \cdots & 0 \\ \vdots & \ddots & \vdots \\ 0 & \cdots & w_m \end{bmatrix} \tag{8.8}$$

Step 3. The positive ideal alternative (the best possible alternative, A^+) is the alternative that has the best value in all attributes and the negative ideal alternative (the worst possible alternative, A^-) is the alternative that has the worst value in all the attributes. Clearly, these alternatives may be virtual and have no external existence. If we assume that J is the index set of positive attributes (profit) and J' is the index set of negative attributes (cost), A^+ and A^- are calculated through formulas (8.9) and (8.10), respectively.

$$A^+ = \left\{ \max_i v_{ij}; j \in J \right\} \cup \left\{ \min_i v_{ij}; j \in J' \right\} = \left(v_1^+, v_2^+, \ldots, v_m^+ \right) \tag{8.9}$$

$$A^- = \left\{ \min_i v_{ij}; j \in J \right\} \cup \left\{ \max_i v_{ij}; j \in J' \right\} = \left(v_1^-, v_2^-, \ldots, v_m^- \right) \tag{8.10}$$

Step 4. At this step, each alternative with its values in different attributes considered to be a point in a space that has dimensions equal to the number of problem attributes (m). Then, the distance of these points to the equivalent points of positive ideal and negative ideal alternatives are calculated in the Euclidean norm through formulas (8.11) and (8.12).

$$d_{i+} = \sqrt{\sum_{j=1}^{m} \left(v_{ij} - v_j^+ \right)^2}; 1 \leq i \leq n \tag{8.11}$$

$$d_{i-} = \sqrt{\sum_{j=1}^{m} \left(v_{ij} - v_j^- \right)^2}; 1 \leq i \leq n \tag{8.12}$$

Step 5. Calculate the relative similarity of each alternative to the ideal alternative through formula (8.13).

$$CL_{i+} = \frac{d_{i-}}{d_{i+} + d_{i-}}; \quad 1 \le i \le n \tag{8.13}$$

Step 6. Rank the alternatives according to CL_{i+}.

Forming the decision matrix and determining the weights of the attributes are among the key steps in every MADM method, including the TOPSIS method. Because it has a direct impact on the final results and accuracy in their calculation will increase the strength of the final results of the method as decision support tools. In the following, these two important steps will be performed by the preferential voting models and the hybrid voting TOPSIS method will be presented. In the voting TOPSIS method, the output of the preferential voting models will be used in the form of a decision matrix and the weights of the attributes in the TOPSIS method. Therefore, the focus of this hybrid method is on the participation of experts' opinions in the form of preferential voting models in forming the decision matrix and determining the weight of attributes. Suppose that n homogeneous alternatives $(A_1, A_2, …, A_n)$ are to be ranked considering m attributes $(C_1, C_2, …, C_m)$. The algorithm of this method is as follows:

Step 1. Prioritize the attributes with the help of a team of experts and form the voting table for the selection process of the attributes. In this process, the goal is to determine the weights of the alternatives with the help of experts through the voting process. Then, using the model (4.6), calculate the optimal value for each attribute. If the optimal value of the model (4.6) for an attribute is equal to 1, you can use the ranking models of Chap. 5 (for example, the cross-efficiency model). Finally, the weight of each attribute is calculated through the formula (8.14).

$$w_j = \frac{Z_j}{\sum\limits_{p=1}^{m} Z_p}; \quad 1 \le j \le m \tag{8.14}$$

Step 2. In this step, the decision matrix is formed through a voting process. For this purpose, with the help of the team of experts formed in the previous step, a voting process is formed to determine the score of each alternative in each attribute. Considering the assumed attribute $C_j (j = 1, 2, …, m)$, the alternatives are prioritized based on this attribute and the voting table is formed. Then the model (4.6) is solved for each alternative and the optimal value is calculated. It is better to use the results of ranking models such as the cross-efficiency method to increase the accuracy of the method at this step as well. Finally, the decision matrix components are calculated through the formula (8.15).

$$r_{ij} = \frac{Z_i}{\sum\limits_{p=1}^{n} Z_p}; \quad 1 \le i \le n \tag{8.15}$$

Step 3. According to the decision matrix and the weights of the attributes obtained in the previous two steps, use the first to sixth steps of the TOPSIS method to rank the alternatives.

In the above hybrid method presented by Soltanifar [27], the accuracy of the TOPSIS method is increased by forming the decision matrix and the weights of the attributes with the help of experts and through voting models. Also, the interaction with DM is also increased through the determination of discrimination intensity functions in the process of solving preferential voting models. Therefore, the hybrid voting TOPSIS method will clearly be a more powerful tool than the TOPSIS method for decision support. Other attempts have been made by researchers to combine TOPSIS and preferential voting models, for example, the method of Ebrahimnejad and Bagherzadeh [28] can be mentioned.

8.4 Voting Linear Assignment Method

Linear Assignment Method (LAM) is one of the MADM methods that uses an integer programming model in the solution process. In this method, only the final priority of the alternatives is determined and not the weights of the alternatives, so the distance between the alternatives is not known. LAM was firstly introduced by Bernardo and Blin [29] and then applied by other researchers [30]. In LAM, there is no need for high accuracy in decision matrix elements, and only the priority of each alternative in each attribute is sufficient. This can be both a strength and a weakness. The strong point is that inaccuracy in forming the matrix will not affect the final results, and the weak point is that the use of less details can reduce the accuracy of the method. LAM algorithm is as follows.

Suppose that n homogeneous alternatives $(A_1, A_2, ..., A_n)$ are to be ranked considering m attributes $(C_1, C_2, ..., C_m)$ and also assume that the weights of the attributes are in the form of $(w_1, w_2, ..., w_m)$, which is specified by DM or calculated from methods such as Shannon's entropy method. The decision matrix of such a problem is as Eq. (8.6).

Step 1. Normalize matrix D using normalization (8.16).

$$N_D = [n_{ij}]_{n \times m} \& n_{ij} = \frac{r_{ij}}{\sum_{k=1}^{n} r_{kj}} \tag{8.16}$$

Step 2. Based on the elements of matrix N_D, prioritize the alternatives according to each attribute. In this step, it is necessary to pay attention to the profit or cost of the attribute.

Step 3. Form the allocation matrix $\boldsymbol{\gamma}$. $\boldsymbol{\gamma}_{n \times n}$ is a square matrix whose row labels are alternatives and its column labels are alternative ranks. The elements of this matrix are calculated through formula (8.17).

$$\gamma_{pk} = \sum_{\text{alternative } p \text{ has rank equal to } k \text{ in attribute } j} w_j \tag{8.17}$$

Step 4. Calculate the final rank of each alternative from solving the binary model (8.18).

$$Max \sum_{i=1}^{n} \sum_{k=1}^{n} \gamma_{ik} h_{ik}$$

$$s.t.$$

$$\sum_{k=1}^{n} h_{ik} = 1, \quad i = 1, 2, \ldots, n \tag{8.18}$$

$$\sum_{i=1}^{n} h_{ik} = 1, \quad k = 1, 2, \ldots, n$$

$$h_{ik} \in \{0, 1\}, \quad i = 1, 2, \ldots, n; \ k = 1, 2, \ldots, n$$

Soltanifar [31] presented the hybrid VLAM to solve the mentioned shortcoming for LAM. In VLAM, instead of the allocation matrix, a voting matrix is formed and finally the weight of the alternatives is determined through a preferential voting model. VLAM steps are very similar to LAM steps. In step 3 of the LAM, it is assumed that the formed matrix $\boldsymbol{\gamma}$ is a voting matrix. With this assumption, step 4 is modified by solving the voting model (8.19). In this way, instead of determining the priority of alternatives, their final weights are obtained. In addition, there is the flexibility to interact more with DM in determining the discrimination intensity functions.

$$Max \sum_{k=1}^{n} v_{pk} \gamma_{pk}$$

$$s.t.$$

$$\sum_{k=1}^{n} v_{pk} \gamma_{jk} \leq 1, \quad j = 1, \ldots, n \tag{8.19}$$

$$v_{pk} - v_{pk+1} \geq d(k, \varepsilon), \quad k = 1, \ldots, n - 1$$

$$v_{pn} \geq d(n, \varepsilon)$$

Example 8.2 Consider the decision matrix in the form of Table 8.7, where three alternatives are to be evaluated according to four attribute. In this table, the weights of the attributes are also known.

The normalized matrix based on formula (8.16) is in the form of Table 8.8.

Assuming that the attributes are of profit type, the prioritization of alternatives for each attribute is presented in Table 8.9.

Thus, matrix γ based on formula (8.17) will be in the form of Table 8.10.

If Table 8.10 is to be considered as the allocation matrix, by the utilization of LAM the results in Table 8.11, are gained.

Table 8.7 Decision matrix

Alternatives	C1	C2	C3	C4
A1	2	5	1100	452
A2	3	4	1200	458
A3	2.5	3	1300	460
Weights of attributes	0.2	0.3	0.1	0.4

Table 8.8 Normalized decision matrix

Alternatives	C1	C2	C3	C4
A1	0.2667	0.4167	0.3056	0.3299
A2	0.4000	0.3333	0.3333	0.3343
A3	0.3333	0.2500	0.3611	0.3358

Table 8.9 Matrix of priorities of each alternative in relative to each of the attribute

Rank	C1	C2	C3	C4
1th	A2	A1	A3	A3
2nd	A3	A2	A2	A2
3rd	A1	A3	A1	A1

Table 8.10 Matrix γ

Alternatives	Rank 1	Rank 2	Rank 3
A1	0.3	0	0.7
A2	0.2	0.8	0
A3	0.5	0.2	0.3

Table 8.11 Final LAM ranking

Rank 1	Rank 2	Rank 3
A3	A2	A1

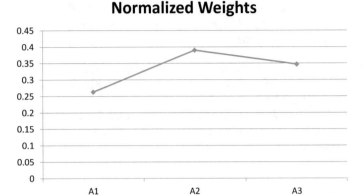

Fig. 8.3 Final VLAM results by $\varepsilon = \varepsilon^* = 0.8451$

Table 8.12 Final VLAM ranking	Rank 1	Rank 2	Rank 3
	A3	A2	A1

Now, we assume that γ is to be taken under consideration as a voting matrix. Thus the final weight of the alternatives is the same as shown in Fig. 8.3. Alternatives can also be easily ranked based on these weights as in Table 8.12.

8.5   Improved KEMIRA Method

KEmeny Median Indicator Ranks Accordance (KEMIRA) is a compensatory MADM method developed by Krylovas et al. [32]. In KEMIRA, the attributes are divided into two or more categories and the ranking of the alternatives is done after determining the priorities and weights of the attributes in two or more different groups by expert judgments. KEMIRA has been used by different researchers in different applications [33–36]. The algorithm of this method is as follows:

Suppose that n homogeneous alternatives $(A_1, A_2, ..., A_n)$ are to be ranked considering m attributes $(C_1, C_2, ..., C_m)$ in the first category and s attributes $(C'_1, C'_2, ..., C'_s)$ in the second category. The decision matrix of such a problem is as Eq. (8.20).

$$D = \begin{bmatrix} x_{11} & \cdots & x_{1m} & y_{11} & \cdots & y_{1s} \\ \vdots & \ddots & \vdots & \vdots & \ddots & \vdots \\ x_{n1} & \cdots & x_{nm} & y_{n1} & \cdots & y_{ns} \end{bmatrix} \tag{8.20}$$

Without losing the generality of the problem, we assume that the attributes are of profit type. If not, we must first convert the related elements to cost attribute, which are under consideration, by inverting the data.

Step 1. Normalize the decision matrix (8.20) by utilizing Eqs. (8.21) and (8.22).

$$\hat{x}_{ji} = \frac{x_{ji} - x_i^-}{x_i^+ - x_i^-}; \quad i = 1, 2, \ldots, m; \, j = 1, 2, \ldots, n. \tag{8.21}$$

$$\hat{y}_{jr} = \frac{y_{jr} - y_r^-}{y_r^+ - y_r^-}; \quad r = 1, 2, \ldots, s; \, j = 1, 2, \ldots, n \tag{8.22}$$

where $x_i^- = \min\limits_{1 \le j \le n} x_{ji}$, $x_i^+ = \max\limits_{1 \le j \le n} x_{ji}$, $i = 1, 2, .., m$, $y_r^- = \min\limits_{1 \le j \le n} y_{jr}$, $y_r^+ = \max\limits_{1 \le j \le n} y_{jr}$, and $r = 1, 2, .., s$.

Step 2. Prioritize first category and second category attribute sets with the help of experts. Let us presume that K experts are utilized for this prioritization; and C_i^k, $(i = 1, 2, \ldots, m; k = 1, 2, \ldots, K)$ is the ith priority attribute of the first category attributes, from the point of view of the kth expert and $C_r'^k$, $(r = 1, 2, \ldots, s; k = 1, 2, \ldots, K)$, is the rth priority attribute of the second category attributes, from the perspective of the kth expert. An attribute with larger value indices has a higher priority than another attribute with lower value indices. The priority matrix of each expert will be determined according to this prioritization for the two attribute categories in the forms of Eqs. (8.23) and (8.24).

$$R_k = \left[a_{ii'}^k\right]_{m \times m}, a_{ii'}^k = \begin{cases} 1 \, if \, C_i^k \succ C_{i'}^k \\ 0 \, if \, C_i^k \prec C_{i'}^k \end{cases}; \quad i, i' = 1, 2, \ldots, m; \, k = 1, 2, \ldots, K \tag{8.23}$$

$$R_k' = \left[a_{rr'}'^k\right]_{s \times s}, a_{rr'}'^k = \begin{cases} 1 \, if \, C_r'^k \succ C_{r'}'^k \\ 0 \, if \, C_r'^k \prec C_{r'}'^k \end{cases}; \quad r, r' = 1, 2, \ldots, s; \, k = 1, 2, \ldots, K \tag{8.24}$$

Step 3. Determine the distance between the prioritization of each expert with other experts in the two attribute categories using Eqs. (8.25) and (8.26).

$$\rho_k = \sum_{k'=1}^{K} \sum_{i=1}^{m} \sum_{i'=1}^{m} \left|a_{ii'}^k - a_{ii'}^{k'}\right|, \quad k = 1, 2, \ldots K \tag{8.25}$$

$$\rho_k' = \sum_{k'=1}^{K} \sum_{r=1}^{s} \sum_{r'=1}^{s} \left|a_{rr'}'^k - a_{rr'}'^{k'}\right|, \quad k = 1, 2, \ldots K \tag{8.26}$$

Step 4. Using Eqs. (8.27) and (8.28), select the expert prioritization with the minimum distance to other expert prioritizations as the reference point for each attribute category. In other words expert k^* will be the reference point expert for the first attribute category, and expert $k^\dagger$ will be the reference expert for the second attribute category. The prioritization of these experts, known as the "median priority components for each attribute group," will be used to determine the weights of the attributes.

$$\rho_{k^*} = \min_k \rho_k \tag{8.27}$$

$$\rho'_{k^\dagger} = \min_k \rho'_k \tag{8.28}$$

Step 5: Determine the weight of the attributes according to the prioritization of the reference experts so that the total weight is equal to 1. Assuming that each weight has members T and T' in its attribute category. These two sets of weights will be in the form of Eqs. (8.29) and (8.30).

$$V = \left\{ \left(v_1^t, v_2^t, \ldots, v_m^t \right) \right\}_{t=1}^T, \quad \sum_{i=1}^m v_i^t = 1, \ t = 1, 2, \ldots, T \tag{8.29}$$

$$U = \left\{ \left(u_1^{t'}, u_2^{t'}, \ldots, u_s^{t'} \right) \right\}_{t'=1}^{T'}, \quad \sum_{r=1}^s u_r^{t'} = 1, \ t' = 1, 2, \ldots, T' \tag{8.30}$$

Step 6. Calculate the weighted sum of the normalized matrix elements for each alternative in the two attribute categories using Eqs. (8.31) and (8.32).

$$X_j^t = \sum_{i=1}^m v_i^t \hat{x}_{ji}, \quad j = 1, 2, \ldots, n; \ \left(v_1^t, v_2^t, \ldots, v_m^t \right) \in V \tag{8.31}$$

$$Y_j^{t'} = \sum_{r=1}^s u_r^{t'} \hat{y}_{jr}, \quad j = 1, 2, \ldots, n; \ \left(u_1^{t'}, u_2^{t'}, \ldots, u_s^{t'} \right) \in U \tag{8.32}$$

Step 7. Calculate $F\left(X^t, Y^{t'} \right)$ for each set of weights using Eq. (8.33) and then find the set of weights for which Eq. (8.34) holds.

$$F\left(X^t, Y^{t'} \right) = \sum_{j=1}^n \left| X_j^t - Y_j^{t'} \right|, \ \left(v_1^t, v_2^t, \ldots, v_m^t \right) \in V, \ \left(u_1^{t'}, u_2^{t'}, \ldots, u_s^{t'} \right) \in U$$

$$\tag{8.33}$$

$$F\left(X^*, Y^*\right) = \min_{\substack{1 \le t \le T \\ 1 \le t' \le T'}} F\left(X^t, Y^{t'}\right) \tag{8.34}$$

Step 8. Assuming that $\left((v_1^*, v_2^*, \ldots, v_m^*), (u_1^*, u_2^*, \ldots, u_s^*)\right)$ are the optimal weights of Eq. (8.34), calculate the final score of the alternatives from Eq. (8.35). An alternative with a higher score will have a higher ranking ($E_{The\ best\ alternative} = \max\limits_{1 \le j \le n} E_j$).

$$E_j = \sum_{i=1}^{m} v_i^* \hat{x}_{ji} + \sum_{r=1}^{s} u_r^* \hat{y}_{jr}, \quad j = 1, 2, \ldots, n \tag{8.35}$$

Finding the weights of the attributes to be applied in the prioritization of reference experts is done in Step 5. This is done by listing the weights for which the priority of reference experts is established and considering a certain accuracy. This process can sometimes be very time consuming. Also, choosing weights from a discrete set can cause some details to be overlooked. Soltanifar [36] suggested that these weights be calculated in a continuous process by solving the mathematical programing model (8.36). This model is based on the concepts of preferential voting and can reduce the time complexity of problem solving while increasing the accuracy of weight selection.

$$\min \sum_{j=1}^{n} \left(\left| \sum_{i=1}^{m} (v)_i^{k^*} (\hat{x})_{ji}^{k^*} - \sum_{r=1}^{s} (u)_r^{k^\dagger} (\hat{y})_{jr}^{k^\dagger} \right| \right)$$

$$s.t. \ \sum_{i=1}^{m} (v)_i^{k^*} = 1$$

$$\sum_{r=1}^{s} (u)_r^{k^\dagger} = 1 \tag{8.36}$$

$$(v)_i^{k^*} - (v)_{i+1}^{k^*} \ge d^{k^*}(i, \varepsilon), \quad i = 1, 2, \ldots, m - 1$$

$$(v)_m^{k^*} \ge d^{k^*}(m, \varepsilon)$$

$$(u)_r^{k^\dagger} - (u)_{r+1}^{k^\dagger} \ge d^{k^\dagger}\left(r, \varepsilon'\right), \quad r = 1, 2, \ldots, s - 1$$

$$(u)_s^{k^\dagger} \ge d^{k^\dagger}\left(s, \varepsilon'\right)$$

where $(v)_i^{(k^*)}, i = 1, 2, ..., m;\ (u)_r^{(k^\dagger)}, r = 1, 2, ..., s$ are re-indexed weights that are true in prioritizing reference experts. Soltanifar [36] showed that the mathematical programming (8.36) can be converted into a linear programming model (8.37) which calculates the weights considered by the KEMIRA method in a process with less time complexity and of course more accurate.

$$\min \sum_{j=1}^{n} \left(P_j^+ + P_j^- \right)$$

$s.t.$

$$\sum_{i=1}^{m} (v)_i^{k^*} (\hat{x})_{ji}^{k^*} - \sum_{r=1}^{s} (u)_r^{k^\dagger} (\hat{y})_{jr}^{k^\dagger} = P_j^+ - P_j^-, \quad j = 1, 2, \ldots, n$$

$$\sum_{i=1}^{m} (v)_i^{k^*} = 1$$

$$\sum_{r=1}^{s} (u)_r^{k^\dagger} = 1 \qquad\qquad\qquad (8.37)$$

$$(v)_i^{k^*} - (v)_{i+1}^{k^*} \geq d^{k^*}(i, \varepsilon), \quad i = 1, 2, \ldots, m - 1$$

$$(v)_m^{k^*} \geq d^{k^*}(m, \varepsilon)$$

$$(u)_r^{k^\dagger} - (u)_{r+1}^{k^\dagger} \geq d^{k^\dagger}\left(r, \varepsilon'\right), \quad r = 1, 2, \ldots, s - 1$$

$$(u)_s^{k^\dagger} \geq d^{k^\dagger}\left(s, \varepsilon'\right)$$

$$P_j^+, P_j^- \geq 0, \quad j = 1, 2, \ldots, n$$

Thus, by using a preferential voting model, the process of implementing the KEMIRA method will be improved.

8.6 Voting KEMIRA Method

Another improvement made in the KEMIRA method was made by Soltanifar et al. [37]. They called their method voting KEMIRA. In this method, despite the use of voting models in improving steps 5, 6 and 7; Step 2 is also improved by using voting models. In other words, instead of prioritizing the attributes by the experts, the voting table for the attributes is formed by the experts and the priority of the attributes is determined by the preferential voting models. Suppose that n homogeneous alternatives $(A_1, A_2, \ldots, A_n)$ are to be ranked considering m attributes $(C_1, C_2, \ldots, C_m)$ in the first category and s attributes $(C_1', C_2', \ldots, C_s')$ in the second category with the decision matrix as Eq. (8.20). The algorithm of this method is as follows.

Step 1. Normalize the decision matrix (8.20) by utilizing Eqs. (8.21) and (8.22).

Step 2: Suppose $a_{ii'}$, $(i, i' = 1, 2, \ldots m)$ is the sum of the votes of the *ith* attribute in the first category of attributes in the *i'th* position and $a'_{rr'}$, $(r, r' = 1, 2, \ldots s)$ is the sum of the votes of the *rth* attribute in the second category of attributes in the *r'th* position. To determine the weights of the attributes, solve the model (8.38).

$$\min \sum_{j=1}^{n} \left(\left| \sum_{i=1}^{m} \left(\left(\sum_{i'=1}^{m} u_{i'} a_{ii'} \right) \hat{x}_{ji} \right) - \sum_{r=1}^{s} \left(\left(\sum_{r'=1}^{s} v_{r'} a'_{rr'} \right) \hat{y}_{ji} \right) \right| \right)$$

$$\max \sum_{i=1}^{m} \left(\sum_{i'=1}^{m} u_{i'} a_{ii'} \right)$$

$$\max \sum_{r=1}^{s} \left(\sum_{r'=1}^{s} v_{r'} a'_{rr'} \right)$$

$s.t.$

$$\sum_{i'=1}^{m} u_{i'} a_{ii'} \leq 1, \quad i = 1, 2, \ldots, m$$

$$\sum_{r'=1}^{s} v_{r'} a'_{rr'} \leq 1, \quad = 1, 2, \ldots, s$$

$$u_{i'} - u_{i'+1} \geq d\left(i', \varepsilon\right), \quad i' = 1, 2, \ldots, m-1$$

$$u_m \geq d(m, \varepsilon)$$

$$v_{r'} - v_{r'+1} \geq d'\left(r', \varepsilon'\right), \quad r' = 1, 2, \ldots, s-1$$

$$v_s \geq d'\left(s, \varepsilon'\right)$$

(8.38)

In this model, despite the existence of the process of model (8.36), there are also two voting models to determine the priority of the attributes, which are all combined in one model. Soltanifar et al. [37] showed that this model can be converted into a linear multi-objective model (8.39).

$$\min \sum_{j=1}^{n} \left(P_j^+ + P_j^- \right)$$

$$\max \sum_{i=1}^{m} \left(\sum_{i'=1}^{m} u_{i'} a_{ii'} \right)$$

$$\max \sum_{r=1}^{s} \left(\sum_{r'=1}^{s} v_{r'} a'_{rr'} \right)$$

$s.t.$

$$\sum_{i=1}^{m} \left(\left(\sum_{i'=1}^{m} u_{i'} a_{ii'} \right) \hat{x}_{ji} \right) - \sum_{r=1}^{s} \left(\left(\sum_{r'=1}^{s} v_{r'} a'_{rr'} \right) \hat{y}_{ji} \right) = P_j^+ - P_j^-, \quad j = 1, 2, \ldots, n$$

$$\sum_{i'=1}^{m} u_{i'} a_{ii'} \leq 1, \quad i = 1, 2, \ldots, m$$

$$\sum_{r'=1}^{s} v_{r'} a'_{rr'} \leq 1, \quad r = 1, 2, \ldots, s$$

$$u_{i'} - u_{i'+1} \geq d\left(i', \varepsilon \right), \quad i' = 1, 2, \ldots, m - 1$$

$$u_m \geq d(m, \varepsilon)$$

$$v_{r'} - v_{r'+1} \geq d'\left(r', \varepsilon' \right), \quad r' = 1, 2, \ldots, s - 1$$

$$v_s \geq d'(s, \varepsilon')$$

$$P_j^+, P_j^- \geq 0, \quad j = 1, 2, \ldots, n$$

$$\text{(8.39)}$$

Methods such as the conversion of objective function to constraints method, weighting method, absolute priority method, the goal programming method and like can be used to solve the model (8.39). Soltanifar et al. used the goal programming method.

Step 3. Suppose $\left(\left(v_1^*, v_2^*, \ldots, v_m^* \right), \left(u_1^*, u_2^*, \ldots, u_s^* \right) \right)$ is the optimal solution of the model (8.39), calculate the weight of the attributes of the first and second category respectively with the help of Eqs. (8.40) and (8.41) and rank the alternatives based on the final scores obtained from Eq. (8.42).

$$w_i = \sum_{i'=1}^{m} u_{i'}^* a_{ii'}, \quad i = 1, 2, \ldots, m \qquad \text{(8.40)}$$

$$w'_r = \sum_{r'=1}^{s} v^*_r a'_{rr'}, \quad r = 1, 2, \ldots, s \tag{8.41}$$

$$E_j = \sum_{i=1}^{m} w_i \hat{x}_{ji} + \sum_{r=1}^{s} w'_r \hat{y}_{jr}, \quad j = 1, 2, \ldots, n \tag{8.42}$$

Example 8.3 Suppose we want to evaluate seven alternatives by considering seven attributes that are grouped into two categories of four and three. The attributes have been prioritized by five experts in each category as follows. The decision matrix is also in the form of Table 8.13. The benefit or cost type of each attribute is also specified in the table.

Table 8.14 shows the normalized matrix with Eqs. (8.21) and (8.22).

Alinezhad and Khalili [2] used the KEMIRA method to solve this problem. They showed that since, the fourth and fifth experts provide the minimum value obtained as

Table 8.13 The decision matrix

Alternatives	Cost type criterion (C_1)	Cost type criterion (C_2)	Cost type criterion (C_3)	Cost type criterion (C_4)	Benefit type criterion (C'_1)	Benefit type criterion (C'_2)	Benefit type criterion (C'_3)
A_1	1.500	0.600	2.500	1.370	9.260	3188.600	55,269.000
A_2	3.500	1.200	4.500	0.500	8.640	497.500	9327.000
A_3	0.800	0.500	3.000	0.100	6.440	2484.000	50,798.000
A_4	4.800	1.200	1.600	2.000	11.119	2676.000	56,206.000
A_5	5.500	1.000	1.600	0.300	5.900	3291.000	66,807.000
A_6	0.600	0.700	2.000	0.600	6.090	6490.000	132,136.000
A_7	0.300	0.400	2.000	0.600	5.720	5496.700	123,314.000

Table 8.14 The normalized decision matrix

Alternatives	Cost type criterion (C_1)	Cost type criterion (C_2)	Cost type criterion (C_3)	Cost type criterion (C_4)	Benefit type criterion (C'_1)	Benefit type criterion (C'_2)	Benefit type criterion (C'_3)
A_1	0.154	0.500	0.441	0.024	0.647	0.449	0.374
A_2	0.033	0.000	0.000	0.158	0.534	0.000	0.000
A_3	0.339	0.700	0.276	1.000	0.132	0.331	0.338
A_4	0.008	0.000	1.000	0.000	1.000	0.364	0.382
A_5	0.000	0.100	1.000	0.298	0.033	0.466	0.468
A_6	0.471	0.357	0.690	0.123	0.068	1.000	1.000
A_7	1.000	1.000	0.690	0.123	0.000	0.909	0.928

in Eq. (8.27), their priority ranking is accepted as the median priority component for the first category of attributes as $C_1 \succ C_2 \succ C_4 \succ C_3$. Additionally, the first, second, third and fourth experts provided the minimum value as obtained in Eq. (8.28). Therefore, their priority ranking is accepted as the median priority component for the second category of attributes as $C_3' \succ C_1' \succ C_2'$. They provided a set of "23" weights for the first category of attributes and a set of "14" weights for the second category of attributes, thus, with regards to the set of the attribute weights, the status "322" should be examined to determine $F(X^*, Y^*)$ as in Eq. (8.34). Finally, $F(X^*, Y^*) = 1.285$ and the final weights of the attributes are as $((v_1^*, v_2^*, v_3^*, v_4^*), (u_1^*, u_2^*, u_3^*))$ $= ((0.400, 0.200, 0.200, 0.200), (0.200, 0.000, 0.800))$. By applying the remaining steps of the KEMIRA method, the final ranking of alternatives will be as (8.43).

$$A_7 \succ A_6 \succ A_3 \succ A_4 \succ A_1 \succ A_5 \succ A_2 \tag{8.43}$$

By using the improved KEMIRA method for this problem, the results of Table 8.15 for the weights of attributes and Table 8.16 for the scores of alternatives will be obtained. In solving the model (8.37), all discrimination intensity functions are assumed equal to epsilon and the results are obtained for different epsilons.

Finally, if the voting KEMIRA method is used for this problem, the results of Table 8.17 for attributes weights and Table 8.18 for alternatives score are obtained. In the solution of model (8.39) all discrimination intensity functions are assumed equal to epsilon and the results are obtained for different epsilons.

Table 8.15 Weights of attributes by improved KEMIRA method

	(C_1)	(C_2)	(C_3)	(C_4)	(C_1')	(C_2')	(C_3')	$F(X^*, Y^*)$
KEMIRA	0.40000	0.20000	0.20000	0.20000	0.20000	0.00000	0.80000	1.28520
$\varepsilon = 0.000001$	0.25000	0.25000	0.25000	0.25000	0.24219	0.00000	0.75781	1.18944
$\varepsilon = 0.00001$	0.25002	0.25000	0.24998	0.24999	0.24218	0.00001	0.75781	1.18948
$\varepsilon = 0.0001$	0.25015	0.25005	0.24985	0.24995	0.24209	0.00010	0.75781	1.18977
$\varepsilon = 0.001$	0.25150	0.25050	0.24850	0.24950	0.24119	0.00100	0.75781	1.19270
$\varepsilon = 0.01$	0.26500	0.25500	0.23500	0.24500	0.23225	0.01000	0.75775	1.22209
$\varepsilon = 0.05$	0.32500	0.27500	0.17500	0.22500	0.19248	0.05000	0.75752	1.35267
$\varepsilon = 0.07$	0.35500	0.28500	0.14500	0.21500	0.17260	0.07000	0.75740	1.41796
$\varepsilon = 0.08$	0.37000	0.29000	0.13000	0.21000	0.16266	0.08000	0.75734	1.45060
$\varepsilon = 0.09$	0.38500	0.29500	0.11500	0.20500	0.18000	0.09000	0.73000	1.51576
$\varepsilon_{max} = 0.1$	0.40000	0.30000	0.10000	0.20000	0.20000	0.10000	0.70000	1.58410

Table 8.16 Final scores of alternatives by improved KEMIRA method

	A_1	A_2	A_3	A_4	A_5	A_6	A_7
KEMIRA	0.68320	0.15160	0.82760	0.70880	0.66060	1.23600	1.50500
$\varepsilon = 0.000001$	0.71987	0.17708	0.86686	0.78367	0.71215	1.18453	1.40650
$\varepsilon = 0.00001$	0.71986	0.17707	0.86685	0.78365	0.71213	1.18453	1.40651
$\varepsilon = 0.0001$	0.71983	0.17702	0.86687	0.78346	0.71203	1.18460	1.40668
$\varepsilon = 0.001$	0.71948	0.17652	0.86700	0.78155	0.71098	1.18525	1.40831
$\varepsilon = 0.01$	0.71598	0.17148	0.86828	0.76247	0.70046	1.19168	1.42457
$\varepsilon = 0.05$	0.70042	0.14906	0.87397	0.67765	0.65372	1.22028	1.49685
$\varepsilon = 0.07$	0.69264	0.13785	0.87682	0.63525	0.63035	1.23458	1.53299
$\varepsilon = 0.08$	0.68876	0.13225	0.87824	0.61404	0.61866	1.24173	1.55106
$\varepsilon = 0.09$	0.69231	0.14121	0.87404	0.60970	0.59511	1.22345	1.54381
$\varepsilon_{max} = 0.1$	0.69660	0.15160	0.86930	0.60700	0.57040	1.20270	1.53410

Table 8.17 Weights of attributes by voting KEMIRA method

	(C_1)	(C_2)	(C_3)	(C_4)	(C_1')	(C_2')	(C_3')	$F(X^*, Y^*)$
KEMIRA	0.40	0.20	0.20	0.20	0.20	0.00	0.80	1.2852
$\varepsilon = 0.010000$	0.20	0.13	0.06	0.11	0.10	0.10	0.10	0.59135
$\varepsilon = 0.020000$	0.40	0.26	0.12	0.22	0.20	0.20	0.20	1.1827
$\varepsilon = 0.030000$	0.60	0.39	0.18	0.33	0.30	0.30	0.30	1.774
$\varepsilon = 0.040000$	0.80	0.52	0.24	0.44	0.40	0.40	0.40	2.3654
$\varepsilon = 0.050000$	1.00	0.65	0.30	0.55	0.50	0.50	0.50	2.95675
Normalized weight	0.40	0.26	0.12	0.22	0.33	0.33	0.33	1.69931

Table 8.18 Final scores of alternatives by voting KEMIRA method

	A_1	A_2	A_3	A_4	A_5	A_6	A_7
KEMIRA	0.6832	0.1516	0.8276	0.7088	0.6606	1.236	1.505
$\varepsilon = 0.010000$	0.2719	0.07738	0.36546	0.2362	0.20248	0.40234	0.5686
$\varepsilon = 0.020000$	0.5438	0.15476	0.73092	0.4724	0.40496	0.80468	1.1373
$\varepsilon = 0.030000$	0.8157	0.23214	1.09638	0.7086	0.60744	1.20702	1.7059
$\varepsilon = 0.040000$	1.0876	0.30952	1.46184	0.9448	0.80992	1.60936	2.2745
$\varepsilon = 0.050000$	1.3595	0.3869	1.8273	1.181	1.0124	2.0117	2.8432
Normalized weight	0.7398	0.226	0.8377	0.7052	0.5339	1.0804	1.3822

8.7 Summary

MCDM methods have always been used as a decision support tool in decision analysis. The study of these methods led to the observation of defects in them, and the researchers tried to modify and improve the methods. In this chapter, methods that have been modified and improved by preference voting models have been discussed. The mentioned hybrid methods can be taken into consideration in decision analysis due to reducing the complexity of calculations, persuading the decision maker to interact more with her/him and eliminating disadvantages such as inconsistency. Of course, the application of preferential voting models in improving MCDM methods is not limited to the presented cases and other cases can be mentioned such as the method presented by Soltanifar [38]. The authors hope that more application of preferential voting models in improving decision-making methods can provide better hybrid methods for decision support.

References

1. Soltanifar, M.: An investigation of the most common multi-objective optimization methods with propositions for improvement. Decis. Anal. J. **1**, 100005 (2021)
2. Alinezhad, A., Khalili, J.: New methods and applications in multiple attribute decision making (MADM). In: International Series in Operations Research & Management Science. Springer, pp. 205–215 (2019)
3. Saaty, T.L.: The analytic hierarchy process. Mcgraw Hill, New York (1980)
4. Liu, F.H.F., Hai, H.L.: The voting analytic hierarchy process method for selecting supplier. Int. J. Prod. Econ. **97**(3), 308–317 (2005)
5. Soltanifar, M., Lotfi,H.F.: The voting analytic hierarchy process method for discriminating among efficient decision making units in data envelopment analysis. Comp. Indust. Eng. **60**(4), 585–592 (2011)
6. Hadi-Vencheh, A., Niazi-Motlagh, M.: An improved voting analytic hierarchy process–data envelopment analysis methodology for suppliers selection. Int. J. Comput. Integr. Manuf. **24**(3), 189–197 (2011)
7. Tavana, M., Soltanifar, M., Santos-Arteaga, F.J.: Analytical hierarchy process: revolution and evolution. Annal. Operat. Res. (2022)
8. Kashian, A.R., Soltanifar, M., Kashian, A.M.: Identifying the priorities of investment in the automotive parts manufacturing industry in Semnan province using VAHP method: Resistance economy-based approach. Econ. Reg. Develop. J. Faculty Econ. Admin. Sci. **26**(18), 221–260 (2020)
9. Pishchulov, G., Trautrims, A., Chesney, T., Gold, S., Schwab, L.: The voting analytic hierarchy process revisited: a revised method with application to sustainable supplier selection. Int. J. Prod. Econ. **211**, 166–179 (2019)
10. Soltanifar, M., Zargar, S.M., Homayounfar, M.: Green supplier selection: a hybrid group voting analytical hierarchy process approach. J. Operat. Res. Its Appl. (Appl. Math.) **19**(2), 113–132 (2022)
11. Rezaei, J.: Best-worst multi-criteria decision-making method. Omega **53**, 49–57 (2015)
12. Ahmadi, H.B., Kusi-Sarpong, S., Rezaei, J.: Assessing the social sustainability of supply chains using best worst method. Resour. Conserv. Recycl. **126**, 99–106 (2017)
13. Delice, E.K., Can, G.F.: A new approach for ergonomic risk assessment integrating KEMIRA, best–worst and MCDM methods. Soft. Comput. **24**(19), 15093–15110 (2020)
14. Liang, F., Brunelli, M., Rezaei, J.: Consistency issues in the best worst method: measurements and thresholds. Omega **96**, 102175 (2020)

15. Rezaei, J.: Best-worst multi-criteria decision-making method: some properties and a linear model. Omega **64**, 126–130 (2016)
16. Rezaei, J., Wang, J., Tavasszy, L.: Linking supplier development to supplier segmentation using best worst method. Expert Syst. Appl. **42**(23), 9152–9164 (2015)
17. Rezaei, J., Nispeling, T., Sarkis, J., Tavasszy, L.: A supplier selection life cycle approach integrating traditional and environmental criteria using the best worst method. J. Clean. Prod. **135**, 577–588 (2016)
18. Keršuliene, V., Zavadskas, E.K., Turskis, Z.: Selection of rational dispute resolution method by applying new step-wise weight assessment ratio analysis (SWARA). J. Bus. Econ. Manag. **11**(2), 243–258 (2010)
19. Bodin, L., Gass, S.I.: Exercises for teaching the analytic hierarchy process. INFORMS Trans. Educ. **4**(2), 1–13 (2004)
20. Saaty, T.L.: The modern science of multicriteria decision making and its practical applications: the AHP/ANP approach. Oper. Res. **61**(5), 1101–1118 (2013)
21. Hwang, C.L., Yoon, K.: Multiple attribute decision making: a state of the art survey.In: Lecture Notes in Economics and Mathematical Systems. Berlin, Springer-Verlog (1981)
22. Soltanifar, M., Shahghobadi, S.: Classifying inputs and outputs in data envelopment analysis based on TOPSIS method and a voting model. Inter. J. Business Anal. (IJBAN) **1**(2), 48–63 (2014)
23. Chakraborty, S.: TOPSIS and modified TOPSIS: a comparative analysis. Decis. Anal. J. **2**, 100021 (2022)
24. Nădăban, S., Dzitac, S., Dzitac, I.: Fuzzy TOPSIS: a general view. Proc. Comp. Sci. **91**, 823–831 (2016)
25. Jahanshahloo, G.R., Lotfi, H.F., Davoodi, A.R.: Extension of TOPSIS for decision-making problems with interval data: interval efficiency. Math. Comp. Model. **49**(5–6), 1137–1142 (2009)
26. Giove, S.: Interval TOPSIS for multicriteria decision making. In Lecture Notes in Computer Science, vol. 2486, pp. 56–63. Heidelberg, Springer, Berlin (2002)
27. Soltanifar, M.: Identify the factors affecting the selection of social media and provide the necessary strategy to improve the status of internal social media. Strat. Manag. Res. **26**(78), 99–122 (2020)
28. Ebrahimnejad, A., Bagherzadeh, M.R.: Data envelopment analysis approach for discriminating efficient candidates in voting systems by considering the priority of voters. Hacettepe J. Math. Stat. **45**(1), 165–180 (2016)
29. Bernardo, J.J., Blin, J.M.: A programming model of consumer choice among multi-attributed brands. J. Cons. Res. **4**(2), 111–118 (1977)
30. Hajiagha, S.H.R., Shahbazi, M., Mahdiraji, H.A., Panahian, H.: A bi-objective score-variance based linear assignment method for group decision making with hesitant fuzzy linguistic term sets. Technol. Econ. Dev. Econ. **24**(3), 1125–1148 (2018)
31. Soltanifar, M.: The voting linear assignment method for determining priority and weights in solving MADM problems. J. Appl. Res. Indust. Eng. **8**(Special Issue), 1–17 (2021)
32. Krylovas, A., Zavadskas, E.K., Kosareva, N., Dadelo, S.: New KEMIRA method for determining criteria priority and weights in solving MCDM problem. Int. J. Inf. Technol. Decis. Mak. **13**(6), 1119–1133 (2014)
33. Krylovas, A., Dadelo, S., Kosareva, N., Zavadskas, E.K.: Entropy-KEMIRA approach for MCDM problem solution in human resources selection task. Int. J. Inf. Technol. Decis. Mak. **16**(05), 1183–1209 (2017)
34. Toktas, P., Can, G.F.: Stochastic KEMIRA-M approach with consistent weightings. Int. J. Inf. Technol. Decis. Mak. **18**(03), 793–831 (2019)
35. Delice, E.K., Can, G.: Correction to: A new approach for ergonomic risk assessment integrating KEMIRA, best–worst and MCDM methods. Soft. Comput. **24**, 15111 (2020)
36. Soltanifar, M.: Improved Kemeny median indicator ranks accordance method. Asia-Pacific J. Operat. Res. (2022)

37. Soltanifar, M., Krylovas, A.A., Kosareva, N.N.: Voting-KEMIRA method for determining criteria priority and weights in solving MADM problems, 25 February 2022, PREPRINT (Version 1) available at Research Square. Soft Computing, vol. Accepted (2023)
38. Soltanifar, M.: A new interval for ranking alternatives in multi attribute decision making problems. J. Appl. Res. Indust. Eng. In Press (2022)

Chapter 9
Preferential Voting Based on the Logic of Uncertainty

Abstract Considering the logic of uncertainty in operations research methods usually makes the results of these methods more justified. In this chapter, the preferential voting model with fuzzy logic is presented. Linguistic term is used in this model. In addition to presenting the application of the presented model, the presented fuzzy preferential voting model will also be used to present a hybrid method of fuzzy multi-attribute decision making.

9.1 Fuzzy Preferential Voting Model

The use of group decision-making plays a decisive role in determining organizational goals. In most decisions related to real-world problems, managers have differences of opinion; moreover, in relation to ranking and selecting alternatives, numerous models and methods are available, often with different results. Preferential voting is a very useful tool in gathering opinions for group decision-making, which was discussed in detail in previous chapters. These models can lead to the improvement of the method in many stages of multi-criteria decision-making methods. Since the logic of uncertainty can always increase the accuracy and flexibility of the method, in this chapter we try to rewrite the preferential voting models based on the logic of uncertainty.

The fuzzy theory presented by Zadeh [1] is one of the uncertainty theories. This theory has been used in various fields until today. Many optimization methods and especially decision-making methods have been rewritten based on this logic and provided grounds for achieving more consistent results. Many of these can be found in the work presented by Kahraman et al. [2]. We can also refer to the comprehensive review provided by Liao et al. [3] in this regard. Belman and Zadeh [4] were among the pioneers who proposed the use of fuzzy data in the objective or objective function and constraints of the optimization problem. In this chapter, using the preliminaries presented in Chap. 3, the classical preferential voting model is rewritten based on fuzzy logic. In Chap. 3, the conversion of qualitative linguistic terms into fuzzy numbers was discussed and suggestions were presented regarding the conversion

Table 9.1 Converting qualitative linguistic term to triangular fuzzy numbers

Fuzzy number $\widetilde{A} = (l, m, u)$	Significance
$(0, 0, 0.1)$	Extremely low
$(0, 0.1, 0.3)$	Very low
$(0.1, 0.3, 0.5)$	Low
$(0.3, 0.5, 0.7)$	Median
$(0.5, 0.7, 0.9)$	High
$(0.7, 0.9, 1)$	Very high
$(0.9, 0.9, 1)$	Extremely high

into triangular fuzzy numbers. This work provides a special opportunity to use fuzzy logic in decision-making methods. Sharafi et al. [5] used Table 9.1 to present their fuzzy preference voting model [6]. Although this is only a suggestion and according to the conditions of the problem, other formats can be used.

In the process of fuzzy preferential voting, instead of allocating or not allocating a vote to a candidate, voters determine the merit of that candidate in each voting preference with a qualitative linguistic term. For example, they state that in their opinion, the merit of the first candidate being placed in the third priority is "High". Then, the linguistic term for each candidate in each voting priority is converted into a fuzzy number using the convertion stated in Table 9.1. In this way, each voter presents a table like Table 9.2 in which the merit of assigning voting preferences to candidates is expressed using appropriate fuzzy numbers.

Where, $\widetilde{x}_{ij}^{r}$ is the fuzzy number allocated to the ith candidate in the jth preference by the rth voter. Finally, the voting table in such a process will be in the form of Table 9.3 where $\widetilde{y}_{ij} \approx \sum_{r=1}^{s} \widetilde{x}_{ij}^{r}$, $(i = 1, 2, \ldots, n; j = 1, 2, \ldots, m)$. In fact, in this table, sum of the fuzzy numbers specifying the merit of each candidate in each priority is considered as the final merit of that candidate being placed in the said priority.

Considering the voting Table 9.3 and considering the policy of preferential voting models, model (9.1) is expressed as a fuzzy preferential voting model. It is a fully fuzzy model whose solution can assign a fuzzy efficiency score to each candidate. This model is presented to evaluate the hypothetical candidate p and must be solved for all candidates.

Table 9.2 Converting qualitative voter's terms into fuzzy numbers

Candidates	First place	Second place	...	mth place
1	$\widetilde{x}_{11}^{r}$	$\widetilde{x}_{12}^{r}$	...	$\widetilde{x}_{1m}^{r}$
2	$\widetilde{x}_{21}^{r}$	$\widetilde{x}_{22}^{r}$	...	$\widetilde{x}_{2m}^{r}$
...	...	...	...	...
n	$\widetilde{x}_{n1}^{r}$	$\widetilde{x}_{n2}^{r}$	...	$\widetilde{x}_{nm}^{r}$

Table 9.3 Voting table with the aggregation of the opinions of voters

Candidates	First place	Second place	...	mth place
1	$\tilde{y}_{11}^r$	$\tilde{y}_{12}^r$	...	$\tilde{y}_{1m}^r$
2	$\tilde{y}_{21}^r$	$\tilde{y}_{22}^r$	...	$\tilde{y}_{2m}^r$
...	...	...	...	...
n	$\tilde{y}_{n1}^r$	$\tilde{y}_{n2}^r$	...	$\tilde{y}_{nm}^r$

$$\tilde{Z}_p \approx Max \sum_{j=1}^{m} \tilde{u}_j^p \otimes \tilde{y}_{pj}$$

$s.t.$

$$\sum_{j=1}^{m} \tilde{u}_j^p \otimes \tilde{y}_{ij} \lesssim \tilde{1} \quad i = 1, 2, \ldots, n \tag{9.1}$$

$$\tilde{u}_j^p \ominus \tilde{u}_{j+1}^p \gtrsim \tilde{d}(j, \varepsilon) \quad j = 1, 2, \ldots, m-1$$

$$\tilde{u}_m^p \gtrsim \tilde{d}(m, \varepsilon)$$

where, $\tilde{d}(., \varepsilon)$ is a discrimination intensity function of the model and is utilized to clarify the difference of voting preferences. Model (9.1) is a Fully Fuzzy Linear Programming (FFLP), which has several approaches to solve this set of problems; and in this context, methods presented by [7–10] and [11] can be mentioned. In this paper, the method employed by Nasseri and Mahmoudi [12] will be utilized. With due attention to the fact that, as model (9.1) is triangular and non-negative as to the entire fuzzy numbers used in it; and in accordance with description given for the fuzzy efficiency by Nasseri and Mahmoudi [12], it has the capacity to be converted to the form of model (9.2).

$$\tilde{Z}_p \approx Max \left(\sum_{j=1}^{m} u_j^{Lp} y_{pj}^L, \sum_{j=1}^{m} u_j^{Mp} y_{pj}^M, \sum_{j=1}^{m} u_j^{Up} y_{pj}^U \right)$$

$s.t.$

$$\left(\sum_{j=1}^{m} u_j^{Lp} y_{ij}^L, \sum_{j=1}^{m} u_j^{Mp} y_{ij}^M, \sum_{j=1}^{m} u_j^{Up} y_{ij}^U \right) \lesssim (1, 1, 1) \quad i = 1, 2, \ldots, n \tag{9.2}$$

$$\left(u_j^{Lp} - u_{j+1}^{Lp}, u_j^{Mp} - u_{j+1}^{Mp}, u_j^{Up} - u_{j+1}^{Up} \right) \gtrsim$$

$$\left(d^L(j, \varepsilon), d^M(j, \varepsilon), d^U(j, \varepsilon) \right) \quad j = 1, 2, \ldots, m-1$$

$$\left(u_m^{Lp}, u_m^{Mp}, u_m^{Up} \right) \gtrsim \left(d^L(m, \varepsilon), d^M(m, \varepsilon), d^U(m, \varepsilon) \right)$$

Finally, model (9.2) can be converted to a linear programming model (9.3), on the basis of the presentation conducted by [12].

$$
Z_p = Max \frac{1}{4} \left(\sum_{j=1}^{m} u_j^{Lp} y_{pj}^L + 2 \sum_{j=1}^{m} u_j^{Mp} y_{pj}^M + \sum_{j=1}^{m} u_j^{Up} y_{pj}^U \right)
$$

$s.t.$

$$
\sum_{j=1}^{m} u_j^{Lp} y_{ij}^L \leq 1 \quad i = 1, 2, \ldots, n
$$

$$
\sum_{j=1}^{m} u_j^{Mp} y_{ij}^M \leq 1 \quad i = 1, 2, \ldots, n
$$

$$
\sum_{j=1}^{m} u_j^{Up} y_{ij}^U \leq 1 \quad i = 1, 2, \ldots, n \tag{9.3}
$$

$$
u_j^{Up} \geq u_j^{Mp} \geq u_j^{Lp} \geq 0 \quad j = 1, 2, \ldots, m
$$

$$
u_j^{Up} - u_{j+1}^{Up} \geq d^U(j, \varepsilon) \quad j = 1, 2, \ldots, m-1
$$

$$
u_j^{Mp} - u_{j+1}^{Mp} \geq d^U(j, \varepsilon) \quad j = 1, 2, \ldots, m-1
$$

$$
u_j^{Lp} - u_{j+1}^{Lp} \geq d^L(j, \varepsilon) \quad j = 1, 2, \ldots, m-1
$$

$$
u_m^{Up} \geq d^U(m, \varepsilon), u_m^{Mp} \geq d^M(m, \varepsilon), u_m^{Lp} \geq d^L(m, \varepsilon)
$$

After solving the model (9.3) for the assumed candidate, a fuzzy efficiency score in the form (9.4) can be obtained.

$$
\tilde{Z}_p \approx \left(Z_p^L, Z_p^M, Z_p^U \right) \approx \left(\sum_{j=1}^{m} u_j^{Lp*} y_{pj}^L, \sum_{j=1}^{m} u_j^{Mp*} y_{pj}^M, \sum_{j=1}^{m} u_j^{Up*} y_{pj}^U \right) \tag{9.4}
$$

In this way, a fuzzy efficiency score will be obtained for each candidate, which can be used as the basis for the ranking of that candidate. It should be noted that the ranking methods presented in Chap. 5 can be rewritten based on the presented fuzzy model, for example, [5] presented the cross-efficiency model using this logic. This logic can significantly increase the accuracy and flexibility of the method and make it a more powerful tool for decision support.

Example 9.1 Let us presume that 4 voters have participated in the voting and from amongst three candidates (*A, B* and *C*) have selected two candidates and the merits of each are expressed as in Table 9.4.

According to Table 9.3 and considering the qualitative language terms in Table 9.4, the voting table will be as Table 9.5.

Table 9.4 Voting results of voters

First voter	1st	2nd	Second voter	1st	2nd
Candidate A	Extremely high	Extremely high	Candidate A	Low	Very low
Candidate B	Low	High	Candidate B	High	High
Candidate C	High	Low	Candidate C	Very high	High
Third Voter	1th	2th	Fourth Voter	1th	2th
Candidate A	High	Very high	Candidate A	Extremely low	Low
Candidate B	Very high	High	Candidate B	Very low	Low
Candidate C	Very low	Low	Candidate C	High	Low

Table 9.5 The voting table

Candidate	1st	2nd
A	(1.7, 2.3, 2.8)	(1.5, 2, 2.5)
B	(1.6, 2.4, 3.2)	(1.3, 2, 2.7)
C	(0.5, 0.8, 1.8)	(1.7, 2.4, 3.1)

Table 9.6 Fuzzy efficiency scores

Candidate	A	B	C
Efficiency	(0.54941, 0.73518, 0.91304)	(0.45675, 0.69997, 0.94319)	(0.43423, 0.62517, 0.9064)

By applying the model (9.1) or especially its linear version, i.e. model (9.3) on the voting table, the results of Table 9.6 are obtained.

Now, by using triangular fuzzy number ranking methods, candidates can be ranked.

What has been stated so far in this chapter can be a beginning for the widespread use of the logic of uncertainty in the preferential voting process. Providing fuzzy versions of ranking methods, providing fuzzy versions of group preferential voting models, providing fuzzy versions of voting in the presence of undesirable voters and models like this can be a good development in this field. In addition, uncertainty logic is not exclusive to fuzzy logic and other areas such as stochastic logic or interval logic can also be used in this regard. In the rest of this chapter, the fuzzy version of a multi-criteria decision making method will be presented.

9.2 Fuzzy KEMIRA Method

In Chap. 8, the KEMIRA method and its improved version and voting version were discussed in detail. In the hybrid versions, the classical preferential voting model and implicit politics in it, was used to improve the performance of the methods and increase the flexibility and accuracy of them. In the following, the fuzzy version of the method will be presented in which the fuzzy preferential voting model is used to improve the method.

The use of fuzzy preferential voting model to improve the KEMIRA method was presented by Soltanifar et al. [13]. While following the improvements made in [14] and [15], they used fuzzy logic to present their model, which led to more logical final results. Suppose that n homogeneous alternatives $(A_1, A_2, ..., A_n)$ are to be ranked considering m attributes $(C_1, C_2, ..., C_m)$ in the first category and s attributes $(C_1', C_2', ..., C_s')$ in the second category. The decision matrix of such a problem is as Eq. (9.5).

$$
\tilde{D} \approx \begin{bmatrix} \tilde{x}_{11} & \cdots & \tilde{x}_{1m} & \tilde{y}_{11} & \cdots & \tilde{y}_{1s} \\ \vdots & \ddots & \vdots & \vdots & \ddots & \vdots \\ \tilde{x}_{n1} & \cdots & \tilde{x}_{nm} & \tilde{y}_{n1} & \cdots & \tilde{y}_{ns} \end{bmatrix} \tag{9.5}
$$

As can be seen, the decision matrix is presented in a fuzzy form after dividing the attributes into two categories. The algorithm of fuzzy KEMIRA method is as follows.

Step 1. Normalize the fuzzy decision matrix (9.5) using Eqs. (9.6) and (9.7).

$$
\tilde{x}_{ji}^* \approx \frac{\tilde{x}_{ji}}{\sum\limits_{k=1}^{n} \tilde{x}_{ki}}; \quad i = 1, 2, \ldots, m; \quad j = 1, 2, \ldots, n \tag{9.6}
$$

$$
\tilde{y}_{jr}^* \approx \frac{\tilde{y}_{jr}}{\sum\limits_{k=1}^{n} \tilde{y}_{kr}}; \quad r = 1, 2, \ldots, s; \quad j = 1, 2, \ldots, n \tag{9.7}
$$

Step 2. Ask the experts to specify the merit of placing each attribute in each position with an appropriate linguistic term. These linguistic terms can be extracted from one of the ranges presented in Chap. 3. Each linguistic term will actually specify a fuzzy number that was discussed in Chap. 3. Suppose $\tilde{a}_{ii'}^k; i, i' = 1, 2, \ldots, m; k = 1, 2, \ldots, K$ is a fuzzy number assigned to attribute i in position i' by expert k and $\tilde{b}_{rr'}^k; r, r' = 1, 2, \ldots, s; k = 1, 2, \ldots, K$ is a fuzzy number assigned to attribute r in position r' by expert k. In this way, the priority matrix of experts will be in the form of Eqs. (9.8) and (9.9).

$$
\tilde{R}_k \approx \left[\tilde{a}_{ii'}^k \right]_{m \times m}; \quad i, i' = 1, 2, \ldots, m; k = 1, 2, \ldots, K \tag{9.8}
$$

$$
\overline{\tilde{R}}_k \approx \left[\tilde{b}_{rr'}^k \right]_{s \times s}; \quad r, r' = 1, 2, \ldots, s; k = 1, 2, \ldots, K \tag{9.9}
$$

Step 3. Calculate the priority distance between each expert and other experts for each attribute category using Eqs. (9.10) and (9.11). Formulas for triangular fuzzy numbers are explained in Chap. 3.

$$\rho_k = \sum_{k'=1}^{K} \sum_{i=1}^{m} \sum_{i'=1}^{m} \left| V\left(\tilde{a}_{ii'}^{k} \ominus \tilde{a}_{ii'}^{k'}\right)\right|; \quad k = 1, 2, \ldots K \tag{9.10}$$

$$\overline{\rho}_k = \sum_{k'=1}^{K} \sum_{r=1}^{s} \sum_{r'=1}^{s} \left| V\left(\tilde{b}_{rr'}^{k} \ominus \tilde{b}_{rr'}^{k'}\right)\right|; \quad k = 1, 2, \ldots K \tag{9.11}$$

Step 4. Specify the prioritization of the experts with the minimum distance to other experts as the reference prioritization. In other words, considering Eqs. (9.12) and (9.13), experts k^* and $k^\dagger$ will be the reference for the first and second categories, respectively. The prioritization of these experts will be the basis for determining the weight of attributes.

$$\rho_{k^*} = \min_k \rho_k \tag{9.12}$$

$$\overline{\rho}_{k^\dagger} = \min_k \overline{\rho}_k \tag{9.13}$$

Step 5. Find the fuzzy efficiency of each attribute in each category based on the opinion of reference experts and through models (9.14) and (9.15). The attributes will be prioritized based on the results of these models.

$$\tilde{Z}_p \approx Max \sum_{i'=1}^{m} \tilde{u}_{i'}^{p} \otimes \tilde{a}_{pi'}^{k^*}; \quad p = 1, 2, \ldots, m$$

$s.t.$

$$\sum_{i'=1}^{m} \tilde{u}_{i'}^{p} \otimes \tilde{a}_{ii'}^{k^*} \preceq \tilde{1}; \quad i = 1, 2, \ldots, m \tag{9.14}$$

$$\tilde{u}_{i'}^{p} \ominus \tilde{u}_{i'+1}^{p} \succeq \tilde{d}(i', \varepsilon); \quad i' = 1, 2, \ldots, m-1$$

$$\tilde{u}_{m}^{p} \succeq \tilde{d}(m, \varepsilon).$$

$$\tilde{Z}_p \approx Max \sum_{r'=1}^{m} \tilde{u}_{r'}^p \otimes \tilde{b}_{pr'}^{k^*}; \quad p = 1, 2, \dots, s$$

s.t.

$$\sum_{r'=1}^{s} \tilde{u}_{r'}^p \otimes \tilde{b}_{rr'}^{k^\dagger} \succeq \tilde{1}; \quad r = 1, 2, \dots, s \tag{9.15}$$

$$\tilde{u}_{r'}^p \ominus \tilde{u}_{r'+1}^p \succeq \tilde{d}(r', \varepsilon); \quad r' = 1, 2, \dots, s - 1$$

$$\tilde{u}_s^p \succeq \tilde{d}(s, \varepsilon).$$

Step 6: Calculate the final weights of the arrributes based on the prioritization of reference experts and KEMIRA policy using the model (9.16). Note that according to KEMIRA's policy, the weights of the attributes are found in such a way that the results of the priority of the alternatives in the two categories of attributes have the least difference. Also, in this method, the selection will be from a continuous and infinite set with fuzzy logic. In model (9.16), $(\tilde{v})_i^{k^*}, i = 1, 2, \dots, m$; $(\tilde{u})_r^{k^\dagger}, r = 1, 2, \dots, s$ represents the re-indexed fuzzy weights for first and second category features based on the judgment of experts k^* and $k^\dagger$.

$$\min \sum_{j=1}^{n} \left(\left| V \left(\sum_{i=1}^{m} \left((\tilde{v})_i^{k^*} \otimes (\tilde{x}^*)_{ji}^{k^*} \right) \ominus \sum_{r=1}^{s} \left((\tilde{u})_r^{k^\dagger} \otimes (\tilde{y}^*)_{jr}^{k^\dagger} \right) \right) \right| \right)$$

s.t.

$$\sum_{i=1}^{m} (\tilde{v})_i^{k^*} \approx \tilde{1}$$

$$\sum_{r=1}^{s} (\tilde{u})_r^{k^\dagger} \approx \tilde{1} \tag{9.16}$$

$$(\tilde{v})_i^{k^*} \ominus (\tilde{v})_{i+1}^{k^*} \succeq \tilde{d}^{k^*}(i, \varepsilon); \quad i = 1, 2, \dots, m - 1$$

$$(\tilde{v})_m^{k^*} \succeq \tilde{d}^{k^*}(m, \varepsilon)$$

$$(\tilde{u})_r^{k^\dagger} \ominus (\tilde{u})_{r+1}^{k^\dagger} \succeq \tilde{d}^{k^\dagger}(r, \varepsilon'); \quad r = 1, 2, \dots, s - 1$$

$$(\tilde{u})_s^{k^\dagger} \succeq \tilde{d}^{k^\dagger}(s, \varepsilon')$$

Step 6. Calculate the final fuzzy score of each alternative through the Eq. (9.17), in which the optimal weights of the model (9.16) are used. The ranking of alternatives will be based on this score.

$$\tilde{E}_j = \left(\sum_{i=1}^{m} (\tilde{v})_i^{k^*} \otimes (\tilde{x}^*)_{ji}^{k^*} \right) \oplus \left(\sum_{r=1}^{s} (\tilde{u})_r^{k^\dagger} \otimes (\tilde{y}^*)_{jr}^{k^\dagger} \right), \quad j = 1, 2, \dots, n \tag{9.17}$$

Example 9.2 In this section, Example (8.1) presented in Chap. 8 is considered in the fuzzy structure. Table 9.7 is the fuzzy decision matrix for choosing the best car, whose attributes are divided into two categories.

Using Eqs. (9.6) and (9.7), the above matrix is normalized in the form of Table 9.8.

In this problem, three experts have been helped. They have specified their judgments regarding the merit of placing each attribute in each position with a linguistic term, which after converting the linguistic term into triangular fuzzy number, the results of Table 9.9 are obtained.

Using Eqs. (9.10) and (9.11), the following results are obtained.

Table 9.7 Fuzzy decision matrix for choosing the best car

Fuzzy decision matrix	Subjective attributes		Objective attributes	
	Prestige	Comfort	MPG (miles pergallon)	Price
Acura TL	(0.6, 0.7, 0.9)	(0.5, 0.7, 0.9)	(0, 0.2, 0.3)	(0, 0.1, 0.2)
Toyota Camry	(0, 0.1, 0.3)	(0.1, 0.3, 0.5)	(0.1, 0.4, 0.5)	(0, 0.2, 0.4)
Honda Civic	(0.2, 0.3, 0.5)	(0, 0.1, 0.4)	(0.3, 0.5, 0.7)	(0.6, 0.8, 0.9)

Table 9.8 Normalized fuzzy decision matrix for choosing the best car

Fuzzy decision matrix	Subjective attributes		Objective attributes	
	Prestige	Comfort	MPG (miles pergallon)	Price
Acura TL	(0.35, 0.64, 1.13)	(0.28, 0.64, 1.5)	(0, 0.19, 0.75)	(0, 0.09, 0.33)
Toyota Camry	(0, 0.09, 0.38)	(0.06, 0.27, 0.83)	(0.07, 0.36, 1.25)	(0, 0.18, 0.67)
Honda Civic	(0.12, 0.27, 0.63)	(0, 0.09, 0.67)	(0.2, 0.45, 1.75)	(0.4, 0.73, 1.5)

Table 9.9 Fuzzy priority matrices for each attribute category

Expert 1	1st	2nd	Expert 2	1st	2nd	Expert 3	1st	2nd
Prestige	(0.7, 0.9, 1)	(0.5, 0.7, 0.9)	Prestige	(0.5, 0.7, 0.9)	(0.7, 0.9, 1)	Prestige	(0.7, 0.9, 1)	(0.7, 0.9, 1)
Comfort	(0.1, 0.3, 0.5)	(0, 0.1, 0.3)	Comfort	(0, 0.1, 0.3)	(0.1, 0.3, 0.5)	Comfort	(0, 0.1, 0.3)	(0, 0.1, 0.3)
Expert 1	1st	2nd	Expert 2	1st	2nd	Expert 3	1st	2nd
MPG	(0.7, 0.9, 1)	(0.5, 0.7, 0.9)	MPG	(0.5, 0.7, 0.9)	(0.7, 0.9, 1)	MPG	(0.9, 1, 1)	(0.7, 0.9, 1)
Price	(0.1, 0.3, 0.5)	(0, 0.1, 0.3)	Price	(0, 0.1, 0.3)	(0.1, 0.3, 0.5)	Price	(0, 0, 0.1)	(0, 0.1, 0.3)

Table 9.10 Final score and the final ranking of alternatives for fuzzy KEMIRA

Alternatives	Fuzzy score	Total integral	Rank
Acura TL	(0.303, 0.758, 1.847)	0.887	2
Toyota Camry	(0.059, 0.455, 1.542)	0.598	3
Honda Civic	(0.373, 0.788, 2.236)	1.016	1

$\rho_{Expert\,1}=1$	$\overline{\rho}_{Expert\,1}=1.2$
$\rho_{Expert\,2}=1$	$\overline{\rho}_{Expert\,2}=1.2$
$\rho_{Expert3}=0.866667$	$\overline{\rho}_{Expert3}=1.066667$

Therefore, based on the opinions of expert 3 in the subjective and objective category which has the minimum value and after solving the models (9.14) and (9.15), the priority of attributes in each category is determined as (Prestige $\prec$ Comfort) and (MPG $\prec$ Price). Finally, by using the results of the model (9.16) and through the Eq. (9.17), the prioritization of alternatives will be obtained in the form of Table 9.10.

9.3   Summary

Usually, the use of uncertainty logic will justify the results of an optimization method. In this chapter, the fuzzy preferential voting model was presented using fuzzy logic. Then the presented model was used to improve one of the MADM methods called KEMIRA method. The presented fuzzy KEMIRA method not only uses the improvements of preferential voting methods to improve the KEMIRA method, but also uses fuzzy logic. The application of the fuzzy preferential voting model is not exclusive to what has been said, and it is certain that other hybrid methods are presented by combining with this method, which will be a stronger tool for decision support.

Appendix

In this section, GAMS software codes for the models described in this chapter are provided.

```
Sets
    R /R1,R2 /
    J /DMU1*DMU3/
    K(J)
;
ALIAS(J,L);
ALIAS(J,H);
Table      YL(J,R)
```

```
             R1        R2
DMU1        1.5       1.7
DMU2        1.3       1.6
DMU3        1.7       0.5;

Table      YM(J,R)
             R1                  R2
DMU1        2                   2.3
DMU2        2                   2.4
DMU3        2.4                 0.8;

Table      YU(J,R)
             R1                  R2
DMU1        2.5                 2.8
DMU2        2.7                 3.2
DMU3        3.1                 1.8;

Variables
     Z;
   Positive Variables
     EPI,P1(J),P2(J),UL(R),UM(R),UU(R);

Parameters
       EPIP,EL,EM,EU,
       ECL(J,L),ECM(J,L),ECU(J,L)
       YLO(R),YMO(R),YUO(R)
       YLH(R),YMH(R),YUH(R)
       YAIL(R),YAIU(R),YAIM(R);

Equations
   Objective,
Const1,Const2,Const3,Const4,Const5,Const6,Const7,Const8,Const9,
              Const10,Const11

   ObjectiveA,
Const1A,Const2A,Const3A,Const4A,Const5A,Const6A,Const7A,Const8A,Const9A,
              Const10A,Const11A,Const12A,Const13A,Const14A;

Objective..
Z=E=(1/4)*(SUM(R,UU(R)*YUO(R))+2*SUM(R,UM(R)*YMO(R))+SUM(R,UL(R)*YLO(R)));

CONST1(J)..       SUM(R,UL(R)*YL(J,R))  =L= 1;
CONST2(J)..       SUM(R,UM(R)*YM(J,R))  =L= 1;
CONST3(J)..       SUM(R,UU(R)*YU(J,R))  =L= 1;

CONST4(R)..    UL(R)   =L=   UM(R)   ;
CONST5(R)..    UM(R)   =L=   UU(R)   ;

CONST6..       UL('R1') =G= UL('R2') + EPIP;
CONST7..       UL('R2') =G= EPIP;

CONST8..       UM('R1') =G= UM('R2') + EPIP;
CONST9..       UM('R2') =G= EPIP;
```

```
CONST10..        UU('R1') =G= UU('R2') + EPIP;
CONST11..        UU('R2') =G= EPIP;

ObjectiveA..                              Z=E=(1/6)*(SUM(R,UU(R)*(YUH(R) -
YAIU(R)))+4*SUM(R,UM(R)*(YMH(R) -         YAIM(R)))+SUM(R,UL(R)*(YLH(R) -
YAIL(R))));

CONST1A(J)..       SUM(R,UL(R)*YL(J,R)) =L= 1;
CONST2A(J)..       SUM(R,UM(R)*YM(J,R)) =L= 1;
CONST3A(J)..       SUM(R,UU(R)*YU(J,R)) =L= 1;

CONST4A(R)..       UL(R)  =L=  UM(R)  ;
CONST5A(R)..       UM(R)  =L=  UU(R)  ;

CONST6A..          UL('R1') =G= UL('R2') + EPIP;
CONST7A..          UL('R2') =G= EPIP;

CONST8A..          UM('R1') =G= UM('R2') + EPIP;
CONST9A..          UM('R2') =G= EPIP;

CONST10A..         UU('R1') =G= UU('R2') + EPIP;
CONST11A..         UU('R2') =G= EPIP;

CONST12A..         SUM(R,UL(R)*YLO(R)) =G= EL;
CONST13A..         SUM(R,UM(R)*YMO(R)) =G= EM;
CONST14A..         SUM(R,UU(R)*YUO(R)) =G= EU;


Model                          A_MODEL                          /Objective,
Const1,Const2,Const3,Const4,Const5,Const6,Const7,Const8,Const9,
               Const10,Const11/;
Model                          B_MODEL                          /ObjectiveA,
Const1A,Const2A,Const3A,Const4A,Const5A,Const6A,Const7A,Const8A,Const9A,
               Const10A,Const11A,Const12A,Const13A,Const14A/;

File FUZZY /Results.txt/;

Puttl FUZZY 'Title ' System.title, @60 'Page ' System.page//;

Put FUZZY ;


EPIP=0.01;
LOOP(R,
        YAIL(R)=SMIN(J,YL(J,R));
        YAIM(R)=SMIN(J,YM(J,R));
        YAIU(R)=SMIN(J,YU(J,R));
);

*ALPHA=0

LOOP(L,
    LOOP(R,YLO(R)=YL(L,R));
    LOOP(R,YMO(R)=YM(L,R));
```

```
LOOP(R,YUO(R)=YU(L,R));
Solve A_MODEL Using LP Maximizing z;
EL=SUM(R,UL.L(R)*YLO(R));
EM=SUM(R,UM.L(R)*YMO(R));
EU=SUM(R,UU.L(R)*YUO(R));
LOOP(J,K(J)=YES);
K(L)=NO;
ECL(L,L)=EL;
ECM(L,L)=EM;
ECU(L,L)=EU;
LOOP(H$K(H),
      LOOP(R,YLH(R)=YL(H,R));
      LOOP(R,YMH(R)=YM(H,R));
      LOOP(R,YUH(R)=YU(H,R));
      Solve B_MODEL Using LP Maximizing z;
      ECL(H,L)=SUM(R,UL.L(R)*YLO(R));
      ECM(H,L)=SUM(R,UM.L(R)*YMO(R));
      ECU(H,L)=SUM(R,UU.L(R)*YUO(R));
   );
);
PUT'CROSS EFFICENCY MATRIX '/;
PUT'EF-L '/;
LOOP(J,
      LOOP(L,
      PUT ECL(L,J):12:5;
      );
PUT/;
);
PUT/;
PUT'EF-M '/;
LOOP(J,
      LOOP(L,
      PUT ECM(L,J):12:5;
      );
PUT/;
);
PUT/;
PUT'EF-U '/;
LOOP(J,
      LOOP(L,
      PUT ECU(L,J):12:5;
      );
PUT/;
);
```

References

1. Zadeh, L.A.: Fuzzy sets. Inf. Control **8**(3), 338–353 (1965)
2. Kahraman, C., Onar, S.C., Oztaysi, B.: Fuzzy Multicriteria DecisionMaking: A Literature Review. Inter. J. Comp. Intell. Syst. **8**(4), 637–666 (2015)
3. Liao, H., Mi, X., Xu, Z.: A survey of decision-making methods with probabilistic linguistic information: bibliometrics, preliminaries, methodologies, applications and future directions. Fuzzy Optim. Decis. Making **19**, 81–134 (2020)
4. Bellman, R.E., Zadeh, L.A.: Decision-making in a fuzzy environment. Manage. Sci. **17**(4), B141–B164 (1970)
5. Sharafi, H., Soltanifar, M., Lotfi, F.H.: Selecting a green supplier utilizing the new fuzzy voting model and the fuzzy combinative distance-based assessment method. EURO J. Decis. Process. **10**, 100010 (2022)

6. Nasseri, S.H., Ebrahimnejad, A., Cao, B.Y.: Fuzzy Linear Programming: Solution Techniques and Applications, 1st edn. Springer (2017)
7. Allahviranloo, T., Lotfi, F.H., Kiasary, M.K., Kiani, N.A., Alizadeh, L.: Solving fully fuzzy linear programming problem by the ranking function. Appl. Math. Sci. **12**(1–4), 19–32 (2008)
8. Lotfi, F.H., Allahviranloo, T., Jondabeh, M.A., Alizadeh, L.: Solving a full fuzzy linear programming using lexicography method and fuzzy approximate solution. Appl. Math. Model. **33**(7), 3151–3156 (2009)
9. Kumar, A., Kaur, J., Singh, P.: A new method for solving fully fuzzy linear programming problems. Appl. Math. Model. **35**(2), 817–823 (2011)
10. Najafi, H.S., Edalatpanah, S.A.: A note on "A new method for solving fully fuzzy linear programming problems." Appl. Math. Model. **37**(14–15), 7865–7867 (2013)
11. Ezzati, R., Khorram, E., Enayati, R.: A new algorithm to solve fully fuzzy linear programming problems using the MOLP problem. Appl. Math. Model. **39**(12), 3183–3193 (2015)
12. Nasseri, S.H., Mahmoudi, F.: A new approach to solve fully fuzzy linear programming problem. J. Appl. Res. Indust. Eng. **6**(2), 139–149 (2019)
13. Soltanifar, M., Tavana, M., Santos-Arteaga, F.J., Charles, V.: A new fuzzy KEMIRA method with an application to innovation park location analysis and selection. J. Oper. Res. Soc. (2023)
14. Soltanifar, M.: Improved Kemeny median indicator ranks accordance method. Asia-Pacific J. Operat. Res. (2022)
15. Soltanifar, M., Krylovas, A.A., Kosareva, N.: Voting-KEMIRA method for determining criteria priority and weights in solving MADM problems, 25 February 2022. PREPRINT (Version 1) available at Research Square. Soft Comp., vol. Accepted, 2023.

Chapter 10
Applications of Preferential Voting in Industry and Society

Abstract Preferential voting models as well as hybrid Multi-Attribute Decision Making methods based on these models are widely used as decision support tools in industry and society. In this chapter, the applications that have been presented so far in the industry and society of these models and methods are briefly stated. It should be noted that the application of preferential voting models and hybrid MADM methods presented based on them are not exclusive to the presented cases and the authors hope that this application will be expanded in the future.

10.1 Introduction

Decision-making is one of the integral components of management and it is manifested in every management task; in determining the policies of the organization, in formulating goals, designing the organization, selecting, evaluating and in all management practices. Decision-making, identification and selection of alternatives are based on the values and preferences of the decision-maker. Decision-making implies that other alternatives should be considered, and in such a case, we want to identify as many of these alternatives as possible and choose the one that includes 2 features: (1) the greatest probability of success or effectiveness to have (2) It is best suited to goals, desires, lifestyle and values. The decision-making process is always present in people's daily lives, and people's success and their good feelings about life depend on the quality of the decisions they make. But this process is specifically used in determining the goals and policies of governments, organizations, factories, etc., and its quality can guarantee their future. In fact, organizational or institutional decisions are decisions made by senior managers and responsible officials of organizations. This type of decision-making affects organizational behavior. In contrast to individual decision-making, there is group decision-making; in this type of decision-making, the organization's decisions are made as a group. In most organizations, institutions and factories, this kind of decision making is used to determine goals and policies. The importance of the results of decisions has led to the creation of a

179
M. Soltanifar et al., *Preferential Voting and Applications: Approaches Based on Data Envelopment Analysis*, Studies in Systems, Decision and Control 471, https://doi.org/10.1007/978-3-031-30403-3_10

science called the science of decision-making and the scientific methods of decision-making are studied and developed in it. One of the methods of group decision-making based on voters' votes is the preferential voting process. One of the methods designed to make decisions on this basis is the methods based on data coverage analysis policy. These methods are used both directly to collect voters' votes in the decision-making process and to improve and develop other decision-making methods and provide new hybrid methods. In this chapter, applied studies conducted in industry and society will be summarized. Certainly, the use of decision-making methods based on preferential voting is not exclusive to the mentioned cases.

10.2   Summary of Applications

In this section, a summary of applied researches based on preferential voting is presented with the mention of their application type in Table 10.1. The application of preferential voting methods is not exclusive to the mentioned cases, and the authors hope that preferential voting models and hybrid methods based on these models will be increasingly noticed by researchers. It should be noted that many researches that have a theoretical aspect and are not directly used in the practical field are not included in this table. These researches can have the necessary potential for application in decision-making in industry and society.

10.3   Further Discussion and Suggestions

In this chapter, a list of researches that have used preferential voting in industry and society was presented. Some of these researches used the decision-making process based on voters' votes and used preferential voting models, and some used hybrid MADM methods to evaluate and rank alternatives, which used preferential voting policy in the design of the methods. Of course, the range of application of voting methods is not limited to what has been said, and in fact, this method or the hybrid methods designed based on it can be used in all areas that require group decision-making. There is also a lot of research that theoretically presents methods based on preference voting models and has not been directly applied to it; which is not included in the above list. Also, the expansion of preferential voting models and hybrid methods based on preferential voting is also a dynamic process and will definitely be developed in the future.

Table 10.1 A summary of applied research on preferential voting

References	Brief description of the research	Application
[1]	In this research, the basic model of preferential voting was considered and how to determine the weights of the model was studied. Then the cross efficiency method was used to rank the efficient candidates. Finally, the proposed method was applied to the selection of R & D projects	Selecting projects to form an R&D program
[2]	This research uses a preferential voting model based on data envelopment analysis to obtain a ranking for each player. In this research, while not only the strength of the players in each tournament is considered, but also the difficulty of the tournament difference is considered	Ranking players
[3]	In this research, the weight restriction in the preferential voting model have been discussed, and then by providing a strong order on the weights, the candidates have been ranked	Choosing a person for the post of technical director
[4]	This research presents a new method for ranking efficient candidates in preferential voting and does not use information about ineffective candidates to discriminate effective candidates	Implementation to a Java Applet
[5]	In this method, using the preferential voting model, a new hybrid method called Voting Analytic Hierarchy Process (VAHP) process has been presented and used to select the supplier	Selecting supplier
[6]	Using the concepts of preferential voting, this research presents a new mathematical model to find the most preferable alternative in group decision making	Selecting the most preferable alternatives
[7]	In this research, the ranking of homogeneous Decision-Making Units (DMUs) has been studied in Data Envelopment Analysis (DEA). First, efficient units were evaluated using different ranking models in DEA, and then the results of the models were aggregated using the VAHP method and the final result was obtained	Rating of bank branches

(continued)

Table 10.1 (continued)

References	Brief description of the research	Application
[8]	This research modifies the cross-efficiency method for ranking DMUs in DEA with the preferential voting model and uses it to rank bank branches	Comparing the efficiency of bank branches
[9]	This research uses the cross-efficiency method in ranking units in DEA. To solve the problem of different optimal solutions, different secondary goal models are used and the results of these models are selected using the preferential voting model	Rating of bank branches
[10]	This research combines Social Choice Theory (SCT) with its voting systems efficiently with AHP, in various contexts of group decision making	Human–Computer Interaction (HCI)
[11]	In this research, a voting model was designed for groups with unequal power levels of members and some ranking models were developed based on the designed model	Faculty appointment
[12]	This research uses the cross-efficiency method to rank units in DEA. In this research, the power of individual appreciation is used in developing a method that combines cross-evaluation, preference voting, and Ordered Weighted Averaging (OWA)	Ranking baseball players
[13]	This research is an application of the preferential voting model in prioritizing sustainable development criteria effective on open pit mine design	Prioritizing sustainable development criteria affecting open pit mine design
[14]	In this research, a new preferential aggregation algorithm is developed using complementary slack conditions and discriminant analysis	Determine the most appropriate locations for constructing agro-industries
[15]	This research introduces a Linear Programming (LP) model for dispatching rule selection in the presence of multiple criteria. The multi-criteria dispatching rule selection problem is first transformed into a preference voting system, and then a minimax LP model is introduced to solve the corresponding problem	Dispatching rule selection

(continued)

Table 10.1 (continued)

References	Brief description of the research	Application
[16]	In this research, a new model has been designed for group preferential voting in a group with unequal power of members	Ranking of petrochemical companies
[17]	This research has used the preferential voting process and the concept of commonset of weights. In this research, to increase the power of differentiation among candidates, a new model is proposed to obtain the appropriate value of the differentiation factor	Sustainable supply chain management
[18]	This research presents a new fuzzy preferential voting model and uses the hybrid cross-efficiency method to discriminate between candidates in the designed model	Green supply chain
[19]	This research uses preferential voting in the ranking process	Spectrum-based fault localization
[20]	Using the concept of preferential voting, this research designs a Multi-Attribute Decision-Making (MADM) method that assigns an interval to each alternative, then identifies the best alternative by ranking the intervals	Choosing an excavator model
[21]	In this research, a new Group Voting AHP method is designed	Green Supplier Selection
[22]	In this research, the new voting KEMIRA method for solving MADM problems is presented	Choosing a plan to construction a hospital
[23]	In this research, the new improved KEMIRA method for solving MADM problems is presented	Construction management
[24]	In this research, the concept of undesirable voters is introduced and then a model for solving MADM problems is presented	Choosing the best type of cutting fluid
[25]	In this research, the new Voting Linear Assignment (VLAM) method for solving MADM problems is presented	Choosing an excavator model
[26]	In this research, the new voting TOPSIS method for solving MADM problems is presented	Social Media
[27]	In this research, a new group preferential voting model for employee evaluation is presented	Employee performance evaluation

(continued)

Table 10.1 (continued)

References	Brief description of the research	Application
[28]	This research is an application of the VAHP method in the automotive industry	Automotive parts industry

References

1. Green, R., Doyle, J., Cook, W.: Preference voting and project ranking using DEA and cross-evaluation. Eur. J. Oper. Res. **90**, 461–472 (1996)
2. Cook, W.D., Doyle, J., Green, R., Kress, M.: Ranking players in multiple tournaments. Comput. Oper. Res. **23**(9), 869–880 (1996)
3. Noguchi, H., Ogawa, M., Ishii, H.: The appropriate total ranking method using DEA for multiple categorized purposes. J. Comput. Appl. Math. **146**, 155–166 (2002)
4. Obata, T., Ishii, H.: A method for discriminating efficient candidates with ranked voting data. Eur. J. Oper. Res. **151**, 233–237 (2003)
5. Liu, F.H.F., Hai, H.L.: The voting analytic hierarchy process method for selecting supplier. Int. J. Prod. Econ. **97**(3), 308–317 (2005)
6. Emrouznejad, Z.A.M.L.A., Mustafa, A., Komijan, A.R.: Selecting the most preferable alternatives in a group decision making problem using DEA. Expert Syst. Appl. **36**, 9599–9602 (2009)
7. Soltanifar, M., Lotfi, F.H.: The voting analytic hierarchy process method for discriminating among efficient decision making units in data envelopment analysis. Comp. Indus. Eng. **60**(4), 585–592 (2011)
8. Angiz, M.Z., Mustafa, A., Kamali, M.J.: Cross-ranking of decision making units in data envelopment analysis. Appl. Math. Model. **37**, 398–405 (2013)
9. Soltanifar, M., Shahghobadi, S.: Selecting a benevolent secondary goal model in data envelopment analysis cross-efficiency evaluation by a voting model. Socioecon. Plann. Sci. **47**(1), 65–74 (2013)
10. Srdjevic, B., Pipan, M., Srdjevic, Z., Blagojevic, B., Zoranovic, T.: Virtually combining the analytical hierarchy process and voting methods in order to make group decisions. Univ. Access Inf. Soc. **14**, 231–245 (2015)
11. Ebrahimnejad, A., Tavana, M., Santos-Arteaga, F.J.: An integrated data envelopment analysis and simulation method for group consensus ranking. Math. Comput. Simul. **119**, 1–17 (2016)
12. Oukil, A., Amin, G.R.: Maximum appreciative cross-efficiency in DEA: a new ranking method. Comput. Ind. Eng. **81**, 14–21 (2015)
13. Moradi, G., Osanloo, M.: Prioritizing sustainable development criteria affecting open pit mine design: a mathematical model. Proc. Earth Planet. Sci. **15**, 813–820 (2015)
14. Izadikhah, M., Saen, R.F.: A new preference voting method for sustainable location planning using geographic information system and data envelopment analysis. J. Clean. Produc. **137**, 1347–1367 (2016)
15. Amin, G.R., El-Bouri, A.: A minimax linear programming model for dispatching rule selection. Comput. Ind. Eng. **121**, 27–35 (2018)
16. Sharafi, H., Lotfi, F.H., Jahanshahloo, G., Rostamy-malkhalifeh, M., Soltanifar, M., Razipour-GhalehJough, S.: Ranking of petrochemical companies using preferential voting at unequal levels of voting power through data envelopment analysis. Math. Sci. **13**(3), 287–297 (2019)
17. Solving voting system by data envelopment analysis for assessing sustainability of suppliers. Group Decis. Negot. **28**, 641–669 (2019)
18. Sharafi, H., Soltanifar, M., Lotfi, F.H.: Selecting a green supplier utilizing the new fuzzy voting model and the fuzzy combinative distance-based assessment method. EURO J. Decis. Proces. **10**, 100010 (2022)

19. Majd, A., Vahidi-Asl, M., Khalilian, A., Bagheri, B.: ConsilientSFL: using preferential voting system to generate combinatorial ranking metrics for spectrum-based fault localization. Appl. Intell. **52**, 11068–11088 (2022)

20. Soltanifar, M.: A new interval for ranking alternatives in multi attribute decision making problems. J. Appl. Res. Indust. Eng. (In Press, 2022)

21. Soltanifar, M., Zargar, S.M., Homayounfar, M.: Green supplier selection: a hybrid group voting analytical hierarchy process approach. J. Operat. Res. Its Appl. (Appl. Math.) **19**(2), 113–132 (2022)

22. Soltanifar, M., Krylovas, A.A., Kosareva, N.N.: Voting-KEMIRA method for determining criteria priority and weights in solving MADM problems, 25 February 2022, PREPRINT (Version 1) available at Research Square," *Soft Computing,* vol. Accepted, 2023

23. Soltanifar, M.: Improved Kemeny median indicator ranks accordance method. Asia-Pacific J. Operat. Res. (2022)

24. Soltanifar, M., Sharafi, H.: Preferential voting in the presence of undesirable voters. Inter. J. Appl. Decision Sci. (2022)

25. Soltanifar, M.: The voting linear assignment method for determining priority and weights in solving MADM problems. J. Appl. Res. Indust. Eng. **8**(Special Issue), 1–17 (2021)

26. Soltanifar, M.: Identify the factors affecting the selection of social media and provide the necessary strategy to improve the status of internal social media. Strat. Manag. Res. **26**(78), 99–122 (2020)

27. Soltanifar, M., Heidariyeh, S.A.: Employee performance evaluation using a new preferential voting process. Inno. Manage. Operat. Strat. **1**(3), 202–220 (2020)

28. Kashian, A.R., Soltanifar, M., Kashian, A.M.: Identifying the priorities of investment in the automotive parts manufacturing industry in Semnan province using VAHP method: Resistance economy-based approach. Econ. Reg. Develop. J. Faculty of Econ. Admin. Sci. **26**(18), 221–260 (2020)

Printed in the United States
by Baker & Taylor Publisher Services